Challenges in Plant Disease Detection and Recent Advancements

Edited by Amar Bahadur

Published in London, United Kingdom

Challenges in Plant Disease Detection and Recent Advancements
http://dx.doi.org/10.5772/intechopen.1000441
Edited by Amar Bahadur

Contributors
Aditya Kukreti, Amar Bahadur, Andrea Elizabeth Román Ramos, Apurba Das, Ashok Kumar Chaubey, Carlos Eduardo Aucique-Perez, Daisy Senapoty, Enespa, Fayaz A. Mohiddin, Ganga Devi Sinniah, Hiranya Kr. Deva Nath, Kajol Yadav, Lovely Bharti, Malika Ablagh, Mimoun Mokhtari, Mohamed Achouri, Namburi Karunakar Reddy, Niranjan Mahadevan, Noureddine Chtaina, Pranab Dutta, Pranjal Kr. Kaman, Prem Chandra, Shazia Farooq, Vithanala Shiva Sai Swaroop, Wafaa Mokhtari, Zakir Amin

First published in London, United Kingdom, 2024 by IntechOpen
IntechOpen is the global imprint of INTECHOPEN LIMITED, registered in England and Wales, registration number: 11086078, 167-169 Great Portland Street, London, W1W 5PF, United Kingdom

British Library Cataloguing-in-Publication Data
A catalogue record for this book is available from the British Library

Additional hard and PDF copies can be obtained from orders@intechopen.com

Challenges in Plant Disease Detection and Recent Advancements
Edited by Amar Bahadur
p. cm.
Print ISBN 978-0-85466-143-5
Online ISBN 978-0-85466-142-8
eBook (PDF) ISBN 978-0-85466-144-2

For EU product safety concerns:
IN TECH d.o.o., Prolaz Marije Krucifikse Kozulić 3, 51000 Rijeka, Croatia,
info@intechopen.com or visit our website at intechopen.com.

Meet the editor

Dr. Amar Bahadur serves as an assistant professor in the Department of Plant Pathology at the College of Agriculture, Tripura, Lembucherra, Agartala. He obtained his Ph.D. in mycology and plant pathology from Banaras Hindu University, Varanasi, India, and is a life member of various professional societies in India. Dr. Bahadur has authored 33 research papers in international and national journals, as well as over 30 proceedings and book chapters in edited books, in addition to six edited books and one authored book. He has experience reviewing national and international journals and has served as the principal investigator in a DBT-funded research project and as a co-investigator in NMHS, Uttarakhand, India. He has attended and delivered papers at both national and international seminars, conferences, and symposia, and he has also traveled to Thailand. In 2018, he served as the zonal president of NER (North East Zone) in the Indian Phytopathological Society at IARI, New Delhi, and also led the organization of a national symposium in 2019. Over 17 years, he has dedicated himself to research, teaching, and extension work. He has received numerous awards and accolades, including the Outstanding Scientist Award, Best Senior Scientist, Best Young Scientist, Outstanding Faculty, Outstanding Scientist in Agriculture, Excellence in Teaching, Research Padma Award, Fellows of different academic societies, Indian Teacher Award, and Distinguished Agriculture Scientist Award, among others.

Contents

Preface

Crops must be continuously monitored for diseases from the early stages of growth until they are ready for harvest, as plant diseases caused by various organisms like fungi, bacteria, viruses, phytoplasma, spiroplasma, and nematodes lead to significant agricultural losses. The traditional approach to monitoring plant diseases is time-consuming and involves techniques such as human eye observation, microscopic examination, molecular analysis, serological testing, and microbiological methods for disease diagnosis and detection. The field of nanotechnology and nano-diagnostics is incredibly captivating and rapidly progressing, particularly in the detection of both living and nonliving stressors as well as pathogenic ailments in agricultural produce; these approaches, whether direct or indirect, display a high level of sensitivity and technique specificity. At present, the most encouraging methods involve sensors that detect reflectance, temperature, or fluorescence. Assessing plant health relies on the use of digital photographic images, which are crucial tools in the field of plant pathology. Digital cameras provide a user-friendly and straightforward way to capture RGB digital images, which are utilized for identifying and quantifying diseases. These RGB-color images, comprising red, green, and blue channels, have proven effective in detecting biotic stress in plants. The alterations in reflectance caused by plant pathogens and diseases can be attributed to damage in the leaf structure and chemical makeup of the tissue as the disease progresses. Chlorophyll fluorescence imaging devices that measure photosynthetic electron transfer often use LED or laser light sources as active sensors. Plant researchers utilize chlorophyll fluorescence parameters to measure variations in the photosynthetic performance of plants. The leaf surface displays differences in photosynthetic activity resulting from biological and environmental influences. Fluorescence imaging combined with image analysis methods proves valuable for distinguishing and measuring fungal infections. Advanced technologies now make it possible to closely track plant health and promptly identify diseases using image processing, specifically convolutional neural network (CNN) methods, resulting in precise prediction of plant diseases. Several methods have been utilized to create automatic and semi-automatic plant disease detection systems, which have proven to be quicker, more cost-effective, and more precise than the conventional approach. Recently, artificial intelligence (AI) technologies have been utilized in plant pathology to detect plant irregularities and infestations. Deep learning (DL) and machine learning (ML) technologies have proven to be effective in identifying and examining lesions caused by significant abiotic stresses like drought and have improved the detection and management of crop and plant infestations. In order to effectively detect diseases, it is crucial to integrate meteorological and plant health information, including temperature and humidity, to identify diseases. Plants are impacted by a range of biotic and abiotic stress caused by global warming and environmental pollution, leading to noteworthy decreases in yield. Trichoderma, present in various environments, effectively controls a wide variety of plant diseases in an eco-friendly manner by competing for nutrients, engaging in mycoparasitism, producing antibiotics and hydrolytic enzymes, and promoting induced plant resistance mechanisms. The application of this method can enhance soil nutrient absorption, foster plant development, bolster the plant's resilience to environmental stresses, and diminish the occurrence of plant diseases. The evolution of plant pathogens due to the widespread use of pesticides and climate change presents significant challenges in the modern era.

The revised book titled *Challenges in Plant Disease Detection and Recent Advancements* comprehensively addresses the detection and management of economically significant diseases. Detection of plant pathogens at an early stage is crucial for timely management approaches, as they are responsible for significant losses in yield and quality. The revised book will be beneficial to students, scientists, teachers, extension personnel, and researchers, and the editor would like to acknowledge the valuable contribution and cooperation of the authors of each chapter. I would like to extend my heartfelt thanks to everyone who contributed directly or indirectly to the preparation of the book. I am delighted to extend my heartfelt thanks to IntechOpen Limited, located at Great Portland Street, London, W1W 5PF, UK, for their efficient and prompt publication of this edited book.

Dr. Amar Bahadur
Assistant Professor, Plant Pathology,
College of Agriculture,
Tripura, India

Section 1

Disease Detection and Diagnosis

Chapter 1

Chlorophyll a Fluorescence: A Method of Biotic Stress Detection

Carlos Eduardo Aucique-Perez and Andrea Elizabeth Román Ramos

Abstract

Plant diseases are a major threat to food security, causing drastic alterations in plant metabolism upon infection by pathogens. This often results in decreased biomass accumulation, slowed growth rates, and diminished yield components. Pathogens, through various lifestyles such as biotrophic, necrotrophic, and hemibiotrophic, disrupt photosynthesis, the primary metabolic process, via functional and structural damages. Furthermore, the CO_2 assimilation in plants is severely altered by pathogens regardless of their lifestyles. Photosynthetic determinations allow us to establish a perspective about the physiological impairment caused by pathogens related to alterations in the CO_2 flow from the atmosphere to carboxylation sites, stomatal limitations, and photosynthetic performance of photosystem II (PSII). From the changes in the energy, dissipation is possible to establish the functional status of the photochemistry machinery under stress conditions. For the above, chlorophyll a fluorescence (CF) and CF imaging (CFI) arose as a method highly sensible to determine the damage caused by pathogens in plants. This review shows a practical perspective on CF tools using visual method and rapid fluorescence induction kinetics (OJIP-test), for disease detection associated with plant-pathogen interaction studies from the physiological viewpoint, their implications for plant pathology research, applications for the plant phenotyping field, and biotic stress detection.

Keywords: photosynthesis, pathogen, plant-pathogen interaction, primary metabolism, plant phenotyping

1. Introduction

The photosynthetic process is one of the most important metabolic pathways through which plants transform the sunlight into chemical energy obtaining carbohydrates to be used in other metabolic events [1]. In plants, the sunlight energy may be dissipated by photochemistry (photosynthesis light reactions), chlorophyll fluorescence (photon re-emission), and heat (**Figure 1**) [2]. Metabolic process of photosynthesis has been studied in two stages: (i) light reactions: these processes allow the solar energy contained in photons of sunlight to be processed in the membranes of chloroplasts to produce adenosine triphosphate (ATP) and nicotinamide adenine dinucleotide phosphate (NADPH) and (ii) the Calvin-Benson cycle, which produces

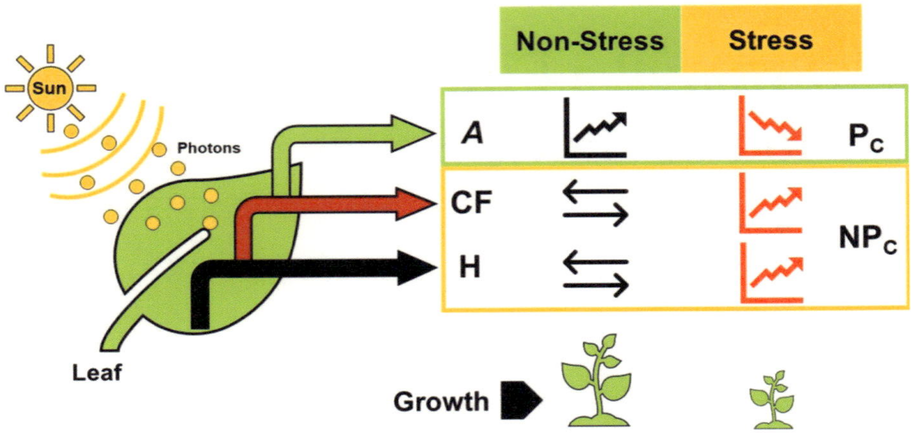

Figure 1.
Energy dissipation pathways (photochemistry (A), chlorophyll fluorescence (CF), and heat (H)) in leaves under non-stress and stress conditions and its effect on plant growth. Photochemical dissipation: P_C; non-photochemical dissipation: NP_C. Design figure credit by Carlos Eduardo Aucique-Perez.

triose-P and, as a result, the recovery of ribulose 1,5-bisphosphate (RuBP), in which how CO_2 is reduced [3, 4]. Chlorophyll fluorescence occurs when the electron transfer from a chlorophyll molecule at the reaction center of the photosystem II (PSII) to a plastoquinone molecule within the plastoquinone transporter, results in fluorescence emission [2]. Finally, the energy cannot be dissipated by photochemistry or fluorescence, and it is dissipated by heat emission [5, 6]. Under this perspective, plants utilize the sun's energy mainly to supply the demands of the primary metabolism; however, external (stresses) and intrinsic (diffusional and biochemical limitations) factors related to the photosynthetic process induce alternative pathways such as fluorescence and heat emission to help in the energy dissipation process to reduce functional and structural damages in the plant cell derivatives of energy accumulation [2].

To study the photosynthesis, several methodologies have been developed involving determinations based on changes in CO_2 and water vapor differential concentrations using infrared sensors (infrared gas analyzer (IRGA)) through which it is possible calculate the leaf gas exchange parameters such as net CO_2 assimilation rate (A), stomatal closed (g_s), CO_2 intercellular concentration (C_i), and transpiration rate (E) [7]. Other methodologies developed in the biochemistry field are commonly applied in photosynthesis studies such as the quantification of the carbohydrates or enzymatic activities of the biochemical components associated with this pathway [8, 9]. Under plant stress, the photosynthetic activity is a main physiological pathway evaluated, as the functionality of the photosynthetic machinery is affected by stress factors [10].

On the other hand, the chlorophyll a fluorescence (CF) has been a non-invasive and non-destructive method successful to study photosynthetic responses from plants under different conditions [11]. Currently, the CF method is applied in diverse fields, especially for rapid diagnosis of abiotic or stress biotic [12, 13] based on changes in the energy dissipation magnitude in the photochemical reactions [10]. Basically, the CF method has been developed from "Kautsky's effect," which was studied when the photosynthetic sample was transferred from the dark to the light, the CF yield increased over a time span of about 1 second. This effect led to increases in the CF yield with a consequent reduction of electron acceptors in the photosynthetic pathway inducing quinone A (Q_A) reduction due to one electron reception which must be transferred to quinone B (Q_B), ensuring the linear electron flow [5]. Therefore, variations in CF characteristics can be indicative of various stresses, including nutrient deficiencies, water deficit, pathogen infections, and high light intensity [2]. By monitoring the CF, scientists, agronomists, and other professionals can gain insights into plant performance and make informed decisions regarding crop management and stress mitigation strategies. In this sense, the CF can be used as a diagnostic tool for assessing plant health and stress responses.

For the plant pathology field, the CF tool has been used for the detection of changes in the energy dissipation pattern in several pathosystems related to fungi, viruses, bacteria, and oomycetes, for example, *Bipolaris sorokiniana* [14], *Fusarium* spp. [11, 15], *Oculimacula acuformis* [11], *Puccinia triticina* [16], *Puccinia coronate* [17], *Bremia lactucae* [18], *Pseudomonas syringae* [19], Grapevine leafroll-associated virus 3 [20, 21], and so on. Advances in the CF instrumentation, including portable devices and remote sensing techniques, have expanded the accessibility and applicability of this technology. In recent years, an increase of portable user-friendly and non-invasive chlorophyll fluorometers had been used for research in the plant pathology field. However, despite the simplicity of the utilization of equipment, the theory and interpretation of data arising from analysis continues to be complex [22].

The use of these devices has allowed crop monitoring on the field and under controlled conditions, plant health management, and ecosystem dynamics [23]. Additionally, the CF imaging technique enable the spatial and temporal visualization of photosynthetic activity, allowing for the assessment of plant performance and the detection of heterogeneity in responses [6]. The integration of the CF with other "omics" technologies further enhances our understanding of photosynthetic regulation and stress responses at the molecular, metabolic, genomic, and phenomic level [24]. Overall, the CF technique plays a crucial role in advancing our knowledge of plant function and resilience. In this sense, the aim of the review was describing the use of the CF as a technique of determination of biotic stress in the plant pathology field.

2. Chlorophyll a fluorescence: method and application

The CF has been used as a technique in photosynthesis research, providing valuable information on the structure and function of the photosynthetic apparatus [10, 12]. In this sense, the photosynthetic response can be estimated by fluorescence changes which study through two types of CF measurement such as continuous excitation and pulsed excitation [5].

Firstly, a method uses pulse amplitude modulation (PAM). The PAM method measures CF induced by high-frequency, modulation pulses (MP) at different stages of the system dynamics driven by saturation pulses (SP) or actinic light (AL) [25, 26]. Thus, the relative yield of fluorescence can be measured in the presence of illumination or can be conducted at field condition [5]. Fundamentally, the CF method determines values associated with basal (F_0 and F_0') and maximal (F_m and F_m') fluorescence on dark and light-adapted leaves from which it has been possible to derive parameters to estimate specific responses of the energy dissipation through the photochemical pathway [5]. Experimentally, F_0, F_0', F_m, and F_m' are obtained from a sequence of light supplying to different intensities which simulated the change from darkness to light adaptation through photosynthetic excitation (**Figure 2**). For obtaining F_0, the dark-adapted leaf is illuminated with weak, modulated measuring pulse

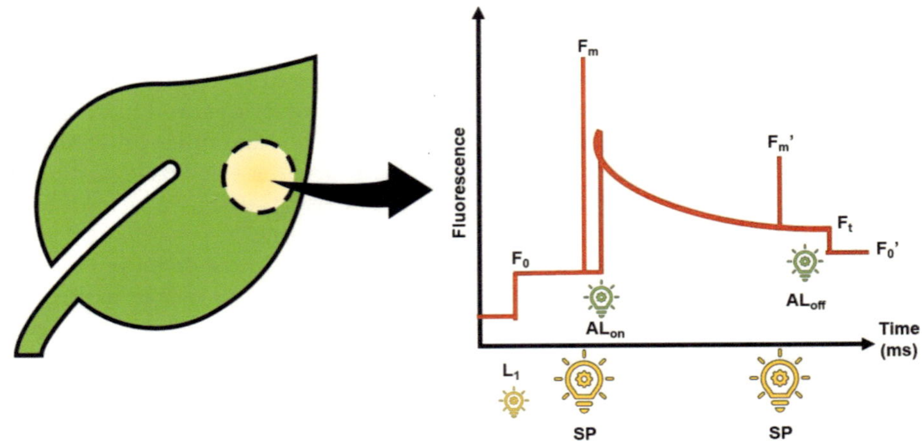

Figure 2.
Fluorescence determination routine. Lowing intensity light (L_1); light saturated pulse (SP); actinic light on (AL_{on}) and off (AL_{off}); basal (F_o) and maximal (F_m) fluorescence in dark-adapted leaf; basal (F_o') and maximal (F_m') fluorescence in light-adapted leaf; fluorescence level immediately before the SP (F_t). Figure adapted from Baker [2].

(L_1 = 0.03 μmol m^{-2} s^{-1}). Immediately, a light saturate pulse (SP = 8000 μmol m^{-2} s^{-1}) is applied for 0.8 s to ensure maximal fluorescence emissions (F_m). Subsequently, on light-adapted leaves, the steady-state fluorescence yield (F_t) is measured following a saturating light pulse (SP = 8000 μmol m^{-2} s^{-1}, 0.8 s) that was applied to achieve the light-adapted maximal fluorescence (F_m'). The actinic light (AL) is then turned off and far-red illumination was applied (2 μmol m^{-2} s^{-1}) to measure the light-adapted basal fluorescence (F_0') [7].

The second method is based on the increase in the CF yield observed when leaves pass through dark to light in a short time period. This induction and fast dynamic changes of CF occur when dark-adapted samples are exposed to light known as Kautsky's effect, as explained above [27], and it is observed as a polyphasic curve, giving information about the efficiency of electron transport through PSII [28, 29]. The CF variations on the OJIP induction curve are also well-known as the OJIP-test. From the O-step to the J-step, it reflects high-order nonlinear dynamics of PSII resulting from light excitation, energy transfer, energy dissipation, and photochemical reactions [25], see **Table 1** for more details. The technique is based on the theory that energy flux through PSII reaction centers can be quantified to provide information about structure and function of the photosynthetic apparatus [33].

Several CF parameters are commonly used to evaluate the performance of the photosynthetic apparatus, such as the maximum quantum yield of photosystem II (PSII), often represented as the F_v/F_m, provides information about the maximum efficiency of PSII photochemistry [5]. The effective quantum yield of PSII (Φ_{PSII}) reflects the actual efficiency of light energy conversion in PSII [2] and non-photochemical quenching (NPQ) that represents the dissipation of excess light energy as heat, thereby protecting the photosynthetic machinery from photodamage [34]. The increases in non-photochemical quenching (measured as NPQ and/or q_N), photoinhibition detected by decreases in F_v/F_m, and inhibition of the electron transport chain can be observed in cases of a stress condition is prolonged [10]. For a higher understanding of the fluorescence parameters, **Table 2** describes the main parameters, their description, calculation, and physiological interpretation.

Technical fluorescence parameters	
Area	Area between fluorescence curve and F_M
M_0	Slope of the curve at the origin of the fluorescence rise. It is a measure of the rate of the primary photochemistry.
V_j	Relative variable fluorescence at 2 ms
V_i	Relative variable fluorescence at 30 ms
$S_M = Area/F_V$	Normalized total complementary area above the OJIP transient (reflecting multiple-turnover Q_A reduction events). It expresses the total electron carriers (EC) per RC.
N	The number of rotations: number of Q_A reduction events between time 0 and t F_M.
Energy fluxes	
ABS	The photon flux absorbed by the antenna of PSII units
TR	The portion of ABS trapped by the active PSII units that leads to Q_A reduction
DI	The portion of ABS dissipated in PSII antenna in process other than trapping
ET	The energy flux associated with the electron transport from Q_A^- to the intersystem electron acceptors
RE	The energy flux associated with the electron transport from Q_A^- to the final electron acceptors of PSI
Quantum yields or flux ratios	
$\phi_{Po} = TR_0/ABS$	Maximum quantum yield of primary photochemistry (at t = 0) $\phi_{Po} = (F_M - F_0)/F_M$
$\psi_{Eo} = ET_0/TR_0$	Probability (at t = 0) that a photon trapped by the PSII reaction center enters the electron transport chain $\psi_{Eo} = 1 - V_j$
$\phi_{Eo} = ET_0/ABS$	Quantum yield of electron transport (at t = 0) $\phi_{Eo} = [(1 - F_0/F_M) \times (1 - V_j)]$
$\phi_{Do} = DI_0/ABS$	Quantum yield (at t = 0) of energy dissipation. $\phi_{Do} = (F_0/F_M)$
$\phi_{Ro} = RE_0/ABS$	Quantum yield of an electron transport from Q_A^- to the final electron acceptors of PSI. $\phi_{Ro} = [(1 - F_0/F_M) \times (1 - V_i)]$
$\delta_{Ro} = RE_0/ET_0$	Efficiency with which an electron from PQH2 is transferred to final PSI acceptors $\delta_{Ro} = (1 - V_i)/(1 - V_j) = (F_M - F)/(F_M - F_j)$
Specific fluxes or specific activities (expressed in arbitrary units)	
ABS/RC	Effective antenna size of an active reaction center (RC). Expresses the total number of photons absorbed by chlorophyll molecules of all RCs divided by the total number of active RCs $ABS/RC = (M_0/V_j)/\phi_{Po}$
TR_0/RC	Maximal trapping rate of PSII. Describes the maximal rate by which an excitation is trapped by the RC $TRo/RC = M_0/V_j$
ET_0/RC	Electron transport in an active RC $ETo/RC = (M_0/V_j)/\psi_{Eo}$
DI_0/RC	Dissipated energy flux per RC $DI_0/RC = ABS/RC - TR_0/RC$
Phenomenological fluxes or phenomenological activities (expressed in arbitrary units)	
ABS/CS_0	Absorption per cross-section (CS) $ABS/CS_0 \approx F_0$ and $ABS/CS_M \approx F_M$

Technical fluorescence parameters	
TR_0/CS	Trapping per cross-section $TR_0/CS = (TR_0/ABS) \times (ABS/CS)$
ET_0/CS	Electron transport per cross-section $ET_0/CS = (ET_0/ABS) \times (ABS/CS)$
RE_0/CS	Electron flux from Q_A^- to final PSI acceptors per cross section of PSI $RE_0/CS = (RE_0/ABS) \times (ABS/CS)$
Performance indexes	
PI_{ABS}	Performance index on adsorption basis $PI_{ABS} = (RC/ABS) \times [\phi_{Po}/(1-\phi_{Po})] \times [\psi_{Eo}/(1-\psi_{Eo})]$
$PI_{ABS,\ total}$	Total performance index on absorption basis $PI_{ABS,\ total} = PI_{ABS} \times [\delta R_0/(1-\delta R_0)]$
PI_{CS}	Performance index on cross section basis $PI_{CS} = PI_{ABS} \times (ABS/CS)$
$PI_{CS,\ total}$	Total performance index on cross section basis $PI_{CS,\ total} = PI_{ABS,\ total} \times (ABS/CS)$

Adapted from reference [30–32].

Table 1.
Parameters and definition related to energy fluxes obtained from OJIP-test.

Parameter	Description	Formula/interpretation
F_0; F_0'	Basal or initial fluorescence from dark and light-adapted leaves, respectively	Level of florescence when all PSII center open (oxidated status)
F_m; F_m'	Maximal fluorescence from dark and light-adapted leaves, respectively	Level of florescence when all PSII center closed (reduced status)
F_v; F_v'	Variable fluorescence from dark and light-adapted leaves, respectively	$F_m - F_0$ It is value associated with the PSII capacity to perform photochemistry
F_v/F_m	Maximum quantum yield of PSII	$(F_m - F_0)/F_m$ Maximum efficiency at which light absorbed by PSII is utilized to reduce Q_A
F_v'/F_m'	Capturing efficiency of the excitation energy by the open PSII reaction centers	$(F_m' - F_0')/F_m'$
q_P	Coefficient for photochemical quenching	$(F_m' - F_t)/(F_m' - F_0')$ Actual photochemical capacity
q_N	Coefficient for photochemical quenching	$1 - [(F_m' - F_0')/F_m - F_0]$ Non-photochemical process activation during the light period
Φ_{PSII}	Actual quantum yield of PSII electron transport	$(F_v'/F_t)/F_m'$
NPQ	Non-photochemical quenching	$(F_m - F_m') - 1$ Apparent heat loss rate from PSII
ETR = J	Rate of electron transport	$PPFD \times \Phi_{PSII} \times f \times \alpha$

PPFD is the photosynthetic photon flux density. f is a factor that accounts for the partitioning of energy between PSII and PSI and is assumed to be 0.5, which indicates that the excitation energy is distributed equally between the two photosystems; and α is the leaf absorbance by the photosynthetic tissues and is assumed to be 0.84.
Adapted from reference [2, 5, 10, 29–31].

Table 2.
Chlorophyll fluorescence parameters used for studies of photosystem II (PSII) photochemistry.

The CF parameters obtained from measurements can provide valuable information about the plant's photosynthetic efficiency, tolerance to stress (abiotic and biotic), or normal physiological condition [6, 35]. Therefore, measuring changes in fluorescence parameters such as F_v/F_m, Φ_{PSII}, and NPQ provides insights into the functionality and regulation of PSII and can indicate early stress detection [34]. In this context, the NPQ represents the protective mechanism employed by plants to dissipate excess absorbed light energy as heat, reducing the potential for photodamage [22]. In addition, the NPQ is influenced by factors such as light intensity, temperature, and the availability of energy sinks within the plant [36]. In practice, the NPQ determination requires leaves' full adaptation to darkness in such a way it reflects the energy dissipation capacity in excess, especially under stress conditions [6]. Recently Tietz et al. [37] proposed the characterization a new set of parameters, $NPQ_{(T)}$, that can be used without dark adaptation and can be rapidly applied during steady-state illumination, potentially even under field conditions.

On the other hand, in the last years, with the developments in optical sensors, the CF imaging (CFI) technique emerged as a powerful tool to study plants under different growth conditions, especially stress factors. In this case, the fluorescence spectral considers the wavelength from UV to infrared, allowing the examination of the spectral region's keys for the plant's physiology responses. In addition, the physical characteristic is related to the emission of fluorescence when the energy is absorbed in the UV region, invisible to the human eye, and posterity is emitted in the visible region and can facilitate obtaining images [38]. Basically, a CFI system utilizes a camera with a long-pass filter, while hindering shorter wavelengths. From the images captured, they are segmented and CF parameters are calculated for each pixel, finishing with the visualization of CF images [39]. With this technique, the obtaining of images from each parameter associated with the CF such as F_v/F_m, Φ_{PSII}, NPQ, and ETR allows the visual exploration of changes in these parameters per area unit [40] and spatial heterogeneity distribution of CF [41]. It has been such a sensibility in this technique that the CFI sensors have been coupled to plant phenotyping systems, obtaining interesting results on indoor and outdoor platforms [10].

3. Photosynthetic alterations by pathogen infection

Under biotic stress, the CO_2 assimilation in plants is severely altered by pathogens infection regardless of their lifestyles (e.g., biotrophic, necrotrophic, and hemibiotrophic) [1, 42]. The biotrophic pathogens get to grow and feed on living plant tissue colonizing plant cells undetected through mechanisms that suppress the plant responses to infection [43]. The biotrophic strategy is considered obligatory parasitism, where the pathogen establishes a molecular and physiological manipulation of the plant cell, and this way it obtains the energy for the growth, such as the causal pathogens associated with rust (e.g., *Hemileia vastatrix*) (**Figure 3A**) and powdery mildew (e.g., *Erysiphales* sp.) diseases [44, 45]. In contrast, the necrotrophic fungi from the first infected plant cell induces rapid cell death using enzymes and toxins to degrade plant tissues, and in this way, the necrotrophic pathogen colonizes the rest of the plant structures [43]. *Botrytis cinerea, Alternaria brassicicola*, and *Neopestalotiopsis* sp. (**Figure 3B**) are species commonly related to the necrotrophic strategy as they have a molecule pool with a high destructive power of the plant's tissues from the first time of the interaction with the host plant [46–48]. Finally, hemibiotrophic fungi have the capacity to combine both biotrophic and necrotrophic strategies [49].

For this strategy, a biotrophic phase allows the colonization of the first plant cell to establish the invading hypha that later deploys other invasive structures in the necrotrophic phase to cause plant cell destruction [50]. An excellent example of the hemibiotrophic strategy is *Pyricularia oryzae* fungi (**Figure 3C**) [51].

During the plant-pathogen interactions, plants infected by pathogens have reported reductions in CO_2 assimilation, stomatal dynamic alterations (stomatal closure), chronic inhibition of the photosystems, higher dissipation of energy by chlorophyll a fluorescence (non-photochemical mechanisms), alterations in the concentration of photosynthetic pigments, chloroplasts structural damage, losses in leaf area index, changes in the source-sink ratio, and changes in the profiles of genes, proteins, and metabolic related to photosynthesis [1, 8, 9, 52]. The pathogen infection in the foliar tissue causes loss in the sunlight conduction capacity due to the cellular damage affecting the photosynthetic activity and consequent development of chlorotic and necrotic zones product of the action of hydrolytic enzymes, non-host selective toxins, and reactive oxygen species (ROS) [1, 53]. While, the plant cell after detection of the pathogen, genes related to photosynthesis are transcript down-regulated to protect the photosynthetic apparatus against the oxidative damage [1, 42].

CO_2 assimilation by the photosynthetic process is a permanent source of primary metabolism, and their products are the main energetic nutrient for pathogens colonizing plant tissues, as well as molecules related to metabolic pathways from secondary metabolism [1, 54]. Studies performed in the last 20 years have allowed understanding how pathogens reprogram the metabolic flux of the carbohydrate's metabolism in detriment of the plant growth and yield demands [1, 8, 42, 54]. From the metabolic perspective, several pathogens rapidly metabolize sucrose through invertase enzymes which hydrolyze sucrose to fructose and glucose increasing the hexoses pool [8, 14, 55]. Metabolites such as fructose, glucose, and hexoses are energetic substrates for other metabolic pathways more complex that under biotic stress those molecules become a nutrient source for pathogens; however, alterations in the sugar concentrations induce the expression of defense genes and cell death considering that sugars act as signaling molecules to regulate the expression of some genes [56]. In the case of foliar pathogens, increases on the invertases activities cause changes in the source-sink relationship dynamic [14, 23, 57].

Figure 3.
Leaf symptoms caused by biotrophic (A; Hemileia vastatrix*), necrotrophic (B;* Neopestalotiopsis *sp.), and hemibiotrophic (C;* Pyricularia oryzae*) fungi in coffee, wheat, and eucalyptus plants, respectively. Picture credits by Carlos Eduardo Aucique-Perez.*

On the other hand, studies have reported that the damage caused by pathogens on leaves led to significant reductions in the net CO_2 assimilation rate (A), stomatal closed (g_s), and transpiration rate (E), as well as alteration in plant water balance and foliar temperature [7, 8, 52, 58–60]. In the case of biotrophic fungi as *Hemileia vastatrix*, *Austropuccinia psidii*, and *Phakopsora pachyrhizi* infecting coffee, eucalyptus, and soybean, respectively, after evidencing symptoms, experiments have demonstrated reductions between 20 and 50% in A, while g_s and E showed higher reductions than A in leaves with severity oscillated between 30 and 50% [45, 61, 62].

Studies focused on energy dissipation through CF have allowed an understanding of the physiological events that happened during the asymptomatic and symptomatic disease phases. It is well-known that the F_v/F_m index is the most extensively utilized chlorophyll a fluorescence parameter stress level determination of abiotic and biotic factors [2]. Under non-stressed conditions, healthy leaves show F_v/F_m values around 0.80–0.83, while inferior F_v/F_m values indicate that a proportion of photosystem II (PSII) reaction centers have been harmed, which induces the photo-inhibitory effect on the photosynthetic machinery and is often led to oxidative stress on plants [63, 64]. In practice, the F_v/F_m parameter is a powerful tool for judging plant photochemical capacity because it acts as a stress indicator for plants; in other words, F_v/F_m measures the operational capacity of the leaf to perform photosynthesis [2, 5, 63, 64].

In the plant pathology field, the use of F_v/F_m has been efficient to measure the pathogen infection damage on the photosynthetic response depending on pathogen lifestyles [10]. However, CF parameters related to photochemical and non-photochemical energy dissipation also have showed interesting findings in plants infected. For instance, in coffee plants with rust symptoms, the values of the yield for dissipation by down-regulation [$Y(NPQ) = (F_s/F_m') - (F_s/F_m)$] and the yield for other non-photochemical (non-regulated) losses [$Y(NO) = F_s/F_m$] were lower and higher, respectively, for the inoculated plants compared to non-inoculated plants, suggesting that *H. vastatrix* after the plant cell colonization reduces the protection capacity of the photosynthetic machinery through antioxidative pathways and inducing a high energy dissipation by heat [65]. In the case of plant virus, high NPQ values showed to be more efficient for the detection of *Potato virus Y* in tobacco plants overproducing endogenous cytokinins [66]. Based on studies performed until now, we can confirm that the CF is a methodology efficient for plant-pathogen interaction studies and with high potential of adaptation in the plant phenotyping approaches.

In the plant resistance field, the photosynthetic damage leads to intrinsic reductions in the defense capacity against the pathogen, and it can reach a higher intensification when abiotic stress factors such as drought, high temperature, or nutritional deficiency/excess participate along the biotic interaction [8, 67, 68]. As known, studies demonstrated that pathogens secret their effectors for linking to receptor specifics in chloroplasts looking for a reduction drastic in the ROS controlling the plant defense mechanisms, suggesting the crucial role of the photosynthetic machinery in the basal plant resistance against pathogen infection [69]. In this sense, the manipulation of photosynthetic activity using photosynthesis inhibitor-specific (DCMU) in wheat cultivars with differential basal resistance to *Pyricularia oryzae* showed increases in the blast severity in both cultivars as a consequence of a low photosynthetic activity affecting the enzymatic responses related to plant resistance such as phenylalanine ammonia-lyase (PAL) and reinforced the concept that photosynthesis was important for wheat resistance to blast [70].

In short, photosynthesis is affected by pathogen infections, and their effects led to reductions in CO_2 assimilation, losses in growth, and crop yield, suggesting that just like abiotic stress, biotic factors threaten food production, as has been reported from the famine in Ireland at XVIII caused by the potato blight epidemic. For the above reasons, deep knowledge about the pathogen strategies to cause disease in plants and how they affect plant metabolism is crucial to designing management alternatives in all fields of agronomy.

4. CF tool to study the plant-pathogen interaction

The use of CF tools in recent years has revealed the utility of this tool from a plant pathology perspective, and CF imaging (CFI) techniques have been used in several plant pathology interactions. However, the equipment for CFI can be expensive, and it is necessary to handle due to the dark adaptation, increasing the time required with a large numbers of plants to test [6]. Today, portable instruments are commercially available being used at field conditions. For instance, CF using the OJIP-test, that is, an induction curve from the O-step to the J-step reflects high-order nonlinear dynamics of PSII resulting from light excitation, energy transfer, energy dissipation, and photochemical reactions (**Table 1**) [33]. As a result, OJIP-test is fast, non-invasive, and sensitive technic [6, 11], becoming an alternative for physiological and plant pathogen studies. Also, this technique allows screening and phenotyping large number of plants (e.g., approximately 500 plants per day) under field conditions, as was reported by Ajigboye et al. [71].

Both techniques offer important information on the PSII damage caused by the infection of pathogens in plants; however, CFI has been more frequently used to detect CF changes in response to foliar diseases in different genotypes due to their visual advantages [8, 40, 72]. In this context, CFI has been used to screen resistant wheat genotypes as well as the early detection of leaf rust caused by *Puccinia triticina* [73]. Kuckenberg et al. [74] observed alteration in F_0 and F_m parameters in asymptomatic wheat leaves affected by powdery mildew and rust. *Ralstonia solanacearum* is a devastating disease for tomato seedling; in a study using CFI, F_v/F_m values between 0.55 and 0.65 were detected, and as a result, the infection was predicted 2 days before visual symptoms [75]. Likewise, CFI was used for the detection of potato virus X (PVX) in *Nicotiana benthamiana* plants; the study demonstrated that Φ_{PSII} and NPQ parameters in infected areas were different significantly starting from the second day after the detection of the virus, considering the CFI as reliable pre-symptomatic detection a viral infection [76]. Also, the CFI was able to detect biotic and abiotic stress in tomato plants, as was reported by Moustaka et al. [77]. In this study, it was able to detect early biotic stress of *Spodoptera exigua* after 15 min of feeding and 30 min after *Botrytis cinerea* inoculation on the onset of water-deficit stress determining as a tool for early stress detection.

On the other hand, F_v/F_m and Y(II) parameters were sensible on coffee leaves inoculated with *H. vastatrix* and fungicide-sprayed ones (**Figure 4**). Those experimental results showed a significant reduction in F_v/F_m and Y(II) values, and they were identified in deep sky blue (F_v/F_m) and yellow-red areas [Y(II)] from 10 days after inoculation with *H. vastatrix*, suggesting that those CF parameters can be used to detect rust symptoms before reaching the incubation period. In coffee plants sprayed with fungicide, F_v/F_m and Y(II) images did not evidence changes in the color patterns and have been similar to control plants (0 days after inoculation), suggesting that the CFI method is an efficient technique for determining photosynthetic alterations caused by *H. vastatrix* as well as effects induced by the fungicide application on the pathogen and the plant, in parallel.

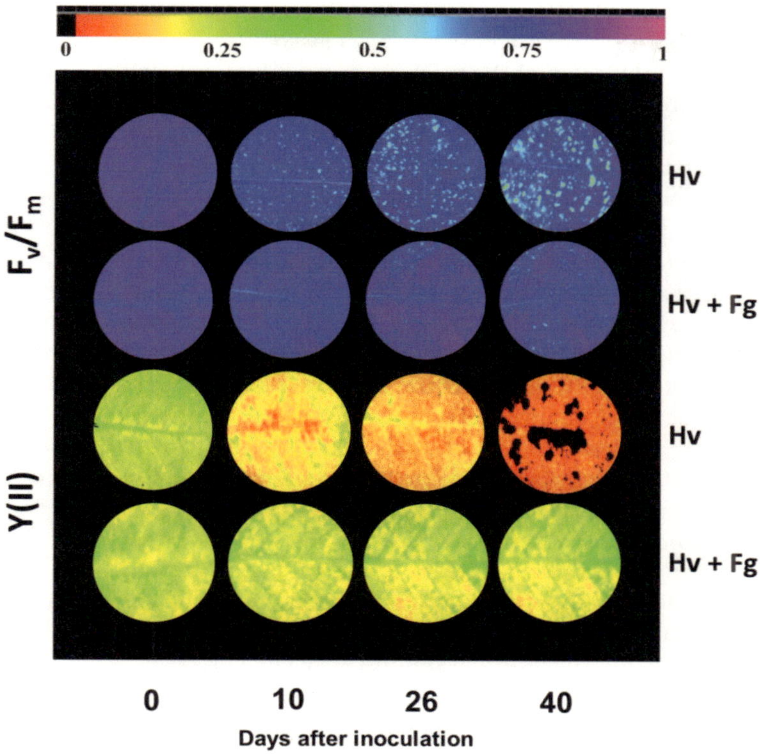

Figure 4.
The maximum quantum yield of PSII (F_v/F_m) and the effective PSII quantum yield [Y(II)] images in coffee plants infected with Hemileia vastatrix *(Hv) sprayed with fungicide (Hv + Fg). The color bar represents the scale from near zero (dark orange) to near one (dark violet). Image supplied by Carlos Eduardo Aucique-Perez.*

The combination of the CFI with other techniques allows an important way to study plant-pathogen interaction. In a study, combining CFI and leaf thermography from thermal infrared (TIR) imaging, among sweet potatoes infected by Sweet potato feathery mottle virus (SPFMV, genus Potyvirus) and Sweet potato chlorotic stunt virus (SPCSV, genus Crinivirus), demonstrated that the F_v/F_m and q_P were the most sensitive parameters for the quantification of virus effects [78]. In oil palms affected with bud rot and lethal wilt, using several CF parameters showed anomalies in the photosynthetic system due to reductions in F_v/F_m and Φ_{PSII} values, as well as increases in the leaf temperature in response to both diseases [79]. At field conditions, a study carried out by Mahlein et al. [80], using a combination of three types of sensors—infrared thermography, CFI, and hyperspectral imaging—symptoms caused by *Fusarium graminearum* and *Fusarium culmorum* were detected successfully. Additionally, the disruption in photosynthetic activity was confirmed through the measurement of F_m in spikelets at 5 dai.

The detection of some diseases in the most cases results in a difficult task due to the complexity between symptomatology observed. In this respect, the combination of CFI and ROS activities detects brown spot and anthracnose in cucumber plants. The difference of the activity levels of the ROS-scavenging enzymes and CFI parameters discriminated cucumber diseases [81]. Likewise, an important disease in citrus such as Citrus Huanglongbing (HLB) had been rapidly detected with high accuracy with the combination of reflectance images and multicolor fluorescence images, which relate to photosynthesis and secondary metabolites [82]. The study demonstrated that the combination of the techniques can effectively detect HLB disease in different infected statuses and cultivars, making it a promising tool for high-throughput HLB disease detection in citrus orchards. Currently, the increases in the interest on artificial intelligent (IA) in the recent years demonstrated the versatility of the CFI for the development of possible tools that discriminate biotic stress. Recently, Sapoukhina et al. [83] proposed a new methodology to develop a data set to generate images from CFI of plants infected by *Pseudomonas syringae* pv. in tomatoes for an automated lesion annotation applied in deep learning, suggesting that IA technology can help to improve the detection levels of diseases and their quantification.

On the other hand, the CF by OJIP-test has been used in several studies to determinate the effects over PSII performance. In this case, Ajigboye et al. [11] using CF determines that existing changes related to the maximum efficiency of PSII photochemistry (F_v'/F_m'), as well as flux of dissipated (DIo/RC), trapped (TRo/RC), or absorbed (ABS/RC) energy per active reaction centers during the attack of necrotrophic pathogens such as *Oculimacula* spp., *F. culmorum*, or *F. avenaceum*. The use of CF measurements and confocal laser-scanning microscopy was used for the detection of *Rosellinia necatrix* that produce white root rot (WRR) disease [84]. This study reveals early stages of WRR disease affect leaf photochemistry before aboveground symptoms appear. Notably, there were significant reductions in photosystem-II trapping efficiency (F_v'/F_m') and the minimal fluorescence yield (F_s/F_0).

In another study, the use of OJIP-test showed that DIo/RC parameter is the best indicator for the diagnosis of *Bursaphelenchus xylophilus* that is a pinewood nematode (PWN) of *Pinus thunbergia* [85], suggesting that this method may assist in disease detection localized in other plant organs, such as roots. In other interaction, it was demonstrated that F_v/F_0, F_0, F_m, and ABS/RC parameters were useful for identifying the physiological changes produced by *F. verticillioides* infection in maize plants [86]. Furthermore, the structural and functional features from OJIP-test was used for the development of a discriminant model for rapid HLB detection [87]. In this study, healthy, asymptomatic, and symptomatic HLB-infected leaves of two different citrus cultivars produced specific changes in the fluorescence patterns. Specifically, HLB-infected symptomatic leaves showed significant reduction in F_v/F_0, Φ_{Eo}, ψ_{Eo}, Φ_{Ro}, Φ_{Po}, PI_{ABS}, PI_{total}, Fm, RE_0/RC and ET_0/RC, and a significant increase in TR_0/RC, F_0, ABS/RC, and DI_0/RC, see **Tables 1** and **2** for details. This research can potentially be used for high-throughput HLB detection when combined with advanced machine learning techniques, offering a promising approach to address the challenges posed by this devastating citrus disease. Overall, these studies have shown that the OJIP-test is a powerful tool to know with higher detail the functional damages caused by pathogens on the photosynthetic machinery.

The CF methodology has been successful in several studies to determine the effect of fungicides on PSII activity during disease control, as mentioned above (**Figure 4**). For instance, the use a fungicide such as isopyrazam and epoxiconazole increased PSII efficiency, ameliorate biomass and grain yield, and reduce disease pressure. Also, the two fungicides increased the efficiency of PSII photochemistry (F_v'/F_m'), as well as improved photosynthetic gas exchange and increased rates of electron transport [71]. In another study, the application of dimethyl sulfoxide (DMSO) in *Solanum lycopersicum* and *Lactuca sativa* plants for the control of *Botrytis cinerea* did not change the chlorophyll fluorescence (F_v/F_m and Φ_{PSII}) [88]. However, one research determined negative alterations on photosynthetic parameters in coffee plants sprayed with tebuconazole and trifloxystrobin, which caused reductions in F_v/F_m, Φ_{PSII}, and ETR, suggesting that there are fungicides that limit the photosynthetic capacity of plants despite disease management [45]. Therefore, the CF methodology contributed to the determination of the physiological effects of fungicides on plants.

Additionally, this tool can be used for phenotyping in breeding programs. In this context, a study conducted by Suarez et al. [89] using CF determination showed that common bean lines induced by biotic stress can be classified as resistant or susceptible phenotypes according to the changes in F_0, F_m, and consequent alteration in F_v/F_m. In barley genotypes infected with *Fusarium culmorum*, alterations in the energy flux associated with the energy dissipation at the PSII ($<DI_0/RC$), limitations in the energy absorption by PSII reaction centers ($<TR_0/RC$), and low energy used for electron transport ($<ET_0/RC$) were determined factors to demonstrate that this response pattern can be used for the selection of barley doubled haploid lines resistance [90]. Those studies indicated the adaptability of CF or CFI techniques associated with plant pathology for the identification of biotic stress, phenotyping, diseases management, and how those technics can be combined with others.

5. Conclusions

The CF is a versatile technique that offers valuable insights into the efficiency and performance of the photosynthetic apparatus. With advancements in methodology and instrumentation, the CF has become an indispensable tool in plant science research, specifically in the plant pathology field.

The measurement setup, parameters obtained, and data interpretation are crucial aspects of the Chl a fluorescence methodology. Researchers employ specialized fluorometers, measure fluorescence parameters such as F_v/F_m, Φ_{PSII}, and NPQ, and interpret the obtained data in conjunction with other physiological measurements for biotic stress detection. However, other parameters have shown high sensitivity depending on the pathosystem, so more studies are necessary for confirming responses and developing new research approaches to improve our knowledge about pathogen infection strategies. Plant physiologists and plant pathologists, as well as plant breeders, hope that, in the next few years, new advances in CF technology will allow a higher understanding of plant responses under pathogen infection, as well as become a powerful tool to support the disease management in issues related to fast detection, evaluation of efficiency in control strategies, and also improving the selection criteria for resistant phenotypes to disease within plant breeding programs.

Acknowledgements

The authors are thankful to Fondo de Investigación para la Agrobiodiversidad, Semillas y Agricultura sustentable (FIASA), grant number FIASA-CA-2023-010.

Conflict of interest

The authors declare no conflict of interest.

Author details

Carlos Eduardo Aucique-Perez[1] and Andrea Elizabeth Román Ramos[2*]

1 Czech Advanced Technology and Research Institute, Palacky University, Olomouc, Czech Republic

2 Agricultural Sciences, Natural Resources and the Environment Faculty, Laboratory of Phytopathology, Bolivar State University, Guaranda, Ecuador

*Address all correspondence to: aroman@ueb.edu.ec

References

[1] Bolton MD. Primary metabolism and plant defense—fuel for the fire. Molecular Plant-Microbe Interactions. 2009;**22**(5):487-497. DOI: 10.1094/MPMI-22-5-0487

[2] Baker NR. Chlorophyll fluorescence: A probe of photosynthesis in vivo. Annual Review of Plant Biology. 2008;**59**:89-113. DOI: 10.1146/annurev.arplant.59.032607.092759

[3] Spreitzer RJ, Salvucci ME. Rubisco: Structure, regulatory interactions, and possibilities for a better enzyme. Annual Review of Plant Biology. 2002;**53**(1):449-475. DOI: 10.1146/annurev.arplant.53.100301.135233

[4] Leister D. Enhancing the light reactions of photosynthesis: Strategies, controversies and perspectives. Molecular Plant. 2023;**14**:4-22. DOI: 10.1016/j.molp.2022.08.005

[5] Maxwell K, Johnson GN. Chlorophyll fluorescence-a practical guide. Journal of Experimental Botany. 2000;**51**(345):659-668. DOI: 10.1093/jexbot/51.345.659

[6] Murchie EH, Lawson T. Chlorophyll fluorescence analysis: A guide to good practice and understanding some new applications. Journal of Experimental Botany. 2013;**64**(13):3983-3998. DOI: 10.1093/jxb/ert208

[7] Aucique-Perez CE, Rodrigues FÁ, Moreira WR, DaMatta FM. Leaf gas exchange and chlorophyll a fluorescence in wheat plants supplied with silicon and infected with *Pyricularia oryzae*. Phytopathology. 2014;**104**(2):143-149. DOI: 10.1094/PHYTO-06-13-0163-R

[8] Aucique-Pérez CE, Rios VS, Neto LBC, Rios JA, Martins SCV, Rodrigues FÁ. Photosynthetic changes in wheat cultivars with contrasting levels of resistance to blast. Journal of Phytopathology. 2020;**168**(11-12):721-729. DOI: 10.1111/jph.12952

[9] Barón M, Flexas J, Delucia EH. Photosynthetic responses to biotic stress. In: Flexas J, Loreto F, Medrano H, editors. Terrestrial Photosynthesis in a Changing Environment. A Molecular, Physiological and Ecological Approach. Cambridge, UK: Cambridge University Press; 2012. pp. 331-350

[10] Pérez-Bueno ML, Pineda M, Barón M. Phenotyping plant responses to biotic stress by chlorophyll fluorescence imaging. Frontiers in Plant Science. 2019;**10**:1135. DOI: 10.3389/fpls.2019.01135

[11] Ajigboye OO, Bousquet L, Murchie EH, Ray RV. Chlorophyll fluorescence parameters allow the rapid detection and differentiation of plant responses in three different wheat pathosystems. Functional Plant Biology. 2016;**43**(4):356-369. DOI: 10.1071/FP15280

[12] Govindjee. Chlorophyll a fluorescence: A bit of basics and history. In: Chlorophyll a fluorescence: A signature of photosynthesis. Dordrecht: Springer; 2004. pp. 1-41. DOI: 10.1007/z978-1-4020-3218-9_1

[13] Schreiber U. Pulse-amplitude-modulation (PAM) fluorometry and saturation pulse method: An overview. Chlorophyll a Fluorescence. 2004;**19**:279-319. DOI: 10.1007/978-1-4020-3218-9_11

[14] Rios JA, Aucique-Pérez CE, Debona D, Cruz Neto LB, Rios VS, Rodrigues FA. Changes in leaf gas

exchange, chlorophyll a fluorescence and antioxidant metabolism within wheat leaves infected by *Bipolaris Sorokiniana*. Annals Applied Biology. 2017;**170**(2):189-203. DOI: 10.1111/aab.12328

[15] Bauriegel E, Herppich WB. Hyperspectral and chlorophyll fluorescence imaging for early detection of plant diseases, with special reference to fusarium spec. infections on wheat. Agriculture. 2014;**4**(1):32-57. DOI: 10.3390/agriculture4010032

[16] Bürling K, Hunsche M, Noga G. Use of blue–green and chlorophyll fluorescence measurements for differentiation between nitrogen deficiency and pathogen infection in winter wheat. Journal of Plant Physiology. 2011;**168**(14):1641-1648. DOI: 10.1016/j.jplph.2011.03.016

[17] Scholes JJD, Rolfe SA. Photosynthesis in localised regions of oat leaves infected with crown rust (*Puccinia Coronata*): Quantitative imaging of chlorophyll fluorescence. Planta. 1996;**199**(4):573-582. DOI: 10.1007/BF00195189

[18] Brabandt H, Bauriegel E, Gärber U, Herppich WB. ΦPSII and NPQ to evaluate *Bremia lactucae*-infection in susceptible and resistant lettuce cultivars. Scientia Horticulturae. 2014;**180**:123-129. DOI: 10.1016/j.scienta.2014.10.012

[19] Rodríguez-Moreno L, Pineda M, Soukupová J, Macho AP, Beuzón CR, Barón M, et al. Early detection of bean infection by *Pseudomonas syringae* in asymptomatic leaf areas using chlorophyll fluorescence imaging. Photosynthesis Research. 2008;**96**(1):27-35. DOI: 10.1007/s11120-007-9278-6

[20] Christov I, Stefanov D, Velinov T, Goltsev V, Georgieva K, Abracheva P, et al. The symptomless leaf infection with grapevine leafroll associated virus 3 in grown in vitro plants as a simple model system for investigation of viral effects on photosynthesis. Journal of Plant Physiology. 2007;**164**(9):1124-1133. DOI: 10.1016/j.jplph.2005.11.016

[21] Montero R, Pérez-Bueno ML, Barón M, Florez-Sarasa I, Tohge T, Fernie AR, et al. Alterations in primary and secondary metabolism in *Vitis vinifera* 'malvasía de Banyalbufar'upon infection with grapevine leafroll-associated virus 3. Physiology Plant. 2016;**157**(4):442-452. DOI: 10.1111/ppl.12440

[22] Guidi L, Lo Piccolo E, Landi M. Chlorophyll fluorescence, photoinhibition and abiotic stress: Does it make any difference the fact to be a C3 or C4 species? Frontiers in Plant Science. 2019;**10**:174. DOI: 10.3389/fpls.2019.00174

[23] Roháček K, Soukupová J, Barták M. Chlorophyll fluorescence: A wonderful tool to study plant physiology and plant stress. In: Schoefs B, editor. Plant cell Compartments- Selected Topics. Kerala India: Research Signpost; 2008. pp. 41-104

[24] Großkinsky DK, Syaifullah SJ, Roitsch T. Integration of multi-omics techniques and physiological phenotyping within a holistic phenomics approach to study senescence in model and crop plants. Journal of Experimental Botany. 2018;**69**(4):825-844. DOI: 10.1093/jxb/erx333

[25] Guo Y, Tan J. Recent advances in the application of chlorophyll a fluorescence from photosystem II. Photochemestry Photobiology. 2015;**91**(1):1-14. DOI: 10.1111/php.12362

[26] Schreiber U. Chlorophyll fluorescence: New instruments for special applications. In: Garab G, editor. Photosynthesis: Mechanisms

and Effects, Proceedings of the XIth International Congress on Photosynthesis; August 17-22; Budapest, Hungary. Vol. I–V. Dordrecht: Springer; 1998. pp. 4253-4258. DOI: 10.1007/978-94-011-3953-3_984

[27] Stirbet A. On the relation between the Kautsky effect (chlorophyll a fluorescence induction) and photosystem II: Basics and applications of the OJIP fluorescence transient. Journal of Photochemistry and Photobiology B: Biology. 2011;**104**(1-2):236-257. DOI: 10.1016/j.jphotobiol.2010.12.010

[28] Strasser RJ, Tsimilli-Michael M, Srivastava A. Analysis of the chlorophyll a fluorescence transient. In: Chlorophyll a Fluorescence. Dordrecht: Springer; 2004. pp. 321-362. DOI: 10.1007/978-1-4020-3218-9_12

[29] Stirbet A, Govindjee. Chlorophyll a fluorescence induction: A personal perspective of the thermal phase, the J–I–P rise. Photosynthesis Research. 2012;**113**:15-61. DOI: 10.1007/s11120-012-9754-5. DOI: 10.1007/s11120-012-9754-5

[30] Strasser RJ, Tsimilli-Michael M, Qiang S, Goltsev V. Simultaneous in vivo recording of prompt and delayed fluorescence and 820-nm reflection changes during drying and after rehydration of the resurrection plant Haberlea rhodopensis. Biochimica Biophysica Acta (BBA)-Bioenergetics. 2010;**1797**:1313-1326. DOI: 10.1016/j.bbabio.2010.03.008

[31] Stirbet A, Lazár D, Kromdijk J, Govindjee. Chlorophyll a fluorescence induction: Can just a one-second measurement be used to quantify abiotic stress responses? Photosynthetica. 2018;**56**:86-104. DOI: 10.1007/s11099-018-0770-3

[32] Bussotti F, Desotgiu R, Pollastrini M, Cascio C. The JIP test: A tool to screen the capacity of plant adaptation to climate change. Scandinavian Journal of Forest Research. 2010;**25**:43-50. DOI: 10.1080/02827581.2010.485777

[33] Strasser RJ, Srivastava A, Tsimilli-Michael M. The fluorescence transient as a tool to characterize and screen photosynthetic samples. Probing Photosynthesis: Mechanism, Regulation & Adaptation. 2000;**25**:445-483

[34] Kalaji HM, Schansker G, Brestic M, Bussotti F, Calatayud A, Ferroni L, et al. Frequently asked questions about chlorophyll fluorescence, the sequel. Photosynthesis Research. 2017;**132**:13-66. DOI: 10.1007/s11120-016-0318-y

[35] Krause GH, Weis E. Chlorophyll fluorescence and photosynthesis: The basics. Annual Review of Plant Physiology and Plant Molecular Biology. 1991;**42**(1):313-349. DOI: 10.1146/annurev.pp.42060191.001525

[36] Nishiyama Y, Murata N. Revised scheme for the mechanism of photoinhibition and its application to enhance the abiotic stress tolerance of the photosynthetic machinery. Applied Microbiology and Biotechnology. 2014;**98**:8777-8796. DOI: 10.1007/s00253-014-6020-0

[37] Tietz S, Hall CC, Cruz JA, Kramer DM. NPQ (T): A chlorophyll fluorescence parameter for rapid estimation and imaging of non-photochemical quenching of excitons in photosystem-II-associated antenna complexes. Plant Cell & Environment. 2017;**40**(8):1243-1255. DOI: 10.1111/pce.12924

[38] Goodwin RH. Fluorescent substances in plants. Annual Review of Plant Physiology. 1953;**4**(1):283-304. DOI: 110.1146/annurev.pp.04.060153.001435

[39] Nedbal L, Whitmarsh J. Chlorophyll fluorescence imaging of leaves and fruits. In: Chlorophyll a Fluorescence: A Signature of Photosynthesis. Dordrecht: Springer; 2004. pp. 389-407. DOI: 10.1007/978-1-4020-3218-9_14

[40] Rolfe SA, Scholes JD. Chlorophyll fluorescence imaging of plant–pathogen interactions. Protoplasma. 2010;**247**(3-4):163-175. DOI: 10.1007/s00709-010-0203-z

[41] Gorbe E, Calatayud A. Applications of chlorophyll fluorescence imaging technique in horticultural research: A review. Scientia Horticulturae (Amsterdam). 2012;**138**:24-35. DOI: 10.1016/j.scienta.2012.02.002

[42] Berger S, Sinha AK, Roitsch T. Plant physiology meets phytopathology: Plant primary metabolism and plant–pathogen interactions. Journal of Experimental Botany. 2007;**58**(15-16):4019-4026. DOI: 10.1093/jxb/erm298

[43] Vleeshouwers VGAA, Oliver RP. Effectors as tools in disease resistance breeding against biotrophic, hemibiotrophic, and necrotrophic plant pathogens. Molecular Plant-Microbe Interactions. 2014;**27**(3):196-206. DOI: 10.1094/MPMI-10-13-0313-IA

[44] Mapuranga J, Zhang N, Zhang L, Chang J, Yang W. Infection strategies and pathogenicity of biotrophic plant fungal pathogens. Frontiers in Microbiology. 2022;**13**:799396. DOI: 10.3389/fmicb.2022.799396

[45] Salcedo-Sarmiento S, Aucique-Pérez CE, Silveira PR, Colmán AA, Silva AL, Mansur PSC, et al. Elucidating the interactions between the rust *Hemileia vastatrix* and a *Calonectria mycoparasite* and the coffee plant. Iscience. 2021;**24**(4):102352

[46] Chen H, He S, Zhang S, Li W, Liu S. The Necrotroph *Botrytis cinerea* BcSpd1 plays a key role in modulating both fungal pathogenic factors and plant disease development. Frontiers in Plant Science. 2022;**13**:820767. DOI: 10.3389/fpls.2022.820767

[47] Belisário R, Aucique-Pérez CE, Abreu LM, Salcedo SS, de Oliveira WM, Furtado GQ. Infection by *Neopestalotiopsis* spp. occurs on unwounded eucalyptus leaves and is favoured by long periods of leaf wetness. Plant Pathology. 2020;**69**(2):194-204. DOI: 10.1111/ppa.13132

[48] Cho Y. How the necrotrophic fungus *Alternaria brassicicola* kills plant cells remains an enigma. Eukaryotic Cell. 2015;**14**(4):335-344. DOI: 10.1128/ec.00226-14

[49] Glazebrook J. Contrasting mechanisms of defense against biotrophic and necrotrophic pathogens. Annual Review of Phytopathology. 2005;**43**:205-227. DOI: 0.1146/annurev.phyto.43.040204.135923

[50] Koeck M, Hardham AR, Dodds PN. The role of effectors of biotrophic and hemibiotrophic fungi in infection. Cell Microbiology. 2011;**13**(12):1849-1857. DOI: 10.1111/j.1462-5822.2011.01665.x

[51] Cruz MFA, Rios JA, Araujo L, Ávila RF. Infection process of *Pyricularia oryzae* on the leaves of wheat seedlings. Tropical Plant Pathology. 2016;**41**:123-127. DOI: 10.1007/s40858-016-0068-6

[52] Aucique-Pérez CE, de Menezes Silva PE, Moreira WR, DaMatta FM, RodriguesFÁ.Photosynthesisimpairments and excitation energy dissipation on wheat plants supplied with silicon and infected with *Pyricularia oryza*e. Plant Physiology and Biochemistry.

2017;**121**:196-205. DOI: 10.1016/j.plaphy.2017.10.023

[53] Mittler R, Vanderauwera S, Suzuki N, Miller G, Tognetti V, Vandepoele K, et al. ROS signaling: The new wave? Trends in Plant Science. 2011;**16**:300-309. DOI: 10.1016/j.tplants.2011.03.007

[54] Bispo WMS, Araujo L, Ávila RT, DaMatta FM, Rodrigues FÁ. Mango resistance against *Ceratocystis fimbriata* is impaired by local starch mobilization. Tropical Plant Pathology. 2016;**41**:225-236. DOI: 10.1007/s40858-016-0092-6

[55] Parker D, Beckmann M, Zubair H, Enot DP, Caracuel-Rios Z, Overy DP, et al. Metabolomic analysis reveals a common pattern of metabolic re-programming during invasion of three host plant species by *Magnaporthe grisea*. The Plant Journal. 2009;**59**(5):723-737. DOI: 10.1111/j.1365-313X.2009.03912.x

[56] Koch KE. Carbohydrate-modulated gene expression in plants. Annual Review of Plant Physiology and Plant Molecular Biology. 1996;**47**(1):509-540. DOI: 10.1146/annurev.arplant.47.1.509

[57] Essmann J, Schmitz-Thom I, Schön H, Sonnewald S, Weis E, Scharte J. RNA interference-mediated repression of cell wall invertase impairs defense in source leaves of tobacco. Plant Physiology. 2008;**147**(3):1288-1299. DOI: 10.1104/pp.108.121418

[58] Ayres PG. Responses of stomata to pathogenic microorganisms. In: Stomatal Physiology. Cambridge, UK: Cambridge University Press; 1981. pp. 205-211

[59] Berdugo CA, Zito R, Paulus S, Mahlein A. Fusion of sensor data for the detection and differentiation of plant diseases in cucumber. Plant Pathology. 2014;**63**(6):1344-1356. DOI: 10.1111/ppa.12219

[60] Bermudez-Cardona MB, Filho JAW, Rodrigues FA. Leaf gas exchange and chlorophyll a fluorescence in maize leaves infected with *Stenocarpella macrospora*. Phytopathology. 2015;**105**(1):26-34. DOI: 10.1094/PHYTO-04-14-0096-R

[61] Alves AA, da Guimarães LM d S, Chaves AR d M, DaMatta FM, Alfenas AC. Leaf gas exchange and chlorophyll a fluorescence of *Eucalyptus urophylla* in response to *Puccinia Psidii* infection. Acta Physiologiae Plantarum. 2011;**33**:1831-1839. DOI: 10.1007/s11738-011-0722-z

[62] Einhardt AM, Ferreira S, Souza GMF, Mochko ACR, Rodrigues FA. Cellular oxidative damage and impairment on the photosynthetic apparatus caused by Asian soybean rust on soybeans are alleviated by nickel. Acta Physiologiae Plantarum. 2020;**42**:1-13. DOI: 10.1007/s11738-020-03108-x

[63] Muller P, Li X-PP, Niyogi KK, Müller P, Li X-PP, Niyogi KK. Non-photochemical quenching. A response to excess light energy. Plant Physiology. 2001;**125**(4):1558-1566. DOI: 10.1104/pp.125.4.1558

[64] Baker NR, Rosenqvist E. Applications of chlorophyll fluorescence can improve crop production strategies: An examination of future possibilities. Journal of Experimental Botany. 2004;**55**(403):1607-1621. DOI: 10.1093/jxb/erh196

[65] Honorato J, Zambolim L, Duarte HSS, Aucique-Pérez CE, Rodrigues FÁ. Effects of epoxiconozale and pyraclostrobin fungicides in the infection process of *Hemileia vastatrix* on coffee leaves as determined by chlorophyll a fluorescence imaging. Journal of Phytopathology. 2015;**163**(11-12):968-977. DOI: 10.1111/jph.12399

[66] Spoustová P, Synková H, Valcke R, Čeřovská N. Chlorophyll a fluorescence as a tool for a study of the potato virus Y effects on photosynthesis of nontransgenic and transgenic P Ssu-Ipt tobacco. Photosynthetica. 2013;**51**(2):191-201. DOI: 10.1007/s11099-013-0023-4

[67] Debona D, Rios JA, Nascimento KJT, Silva LC, Rodrigues FA. Influence of magnesium on physiological responses of wheat infected by *Pyricularia oryzae*. Plant Pathology. 2016;**65**(1):114-123. DOI: 10.1111/ppa.12390

[68] Silva BN, Oliveira LM, Mochko ACR, Chaves JAA, Aucique-Pérez CE, Rodrigues FA. Physiological and biochemical changes in wheat plants infected by *Pyricularia Oryzae* caused by thermal oscillations. Physiological and Molecular Plant Pathology. 2021;**115**:101646. DOI: 10.1016/j.pmpp.2021.101646

[69] de Torres ZM, Littlejohn G, Jayaraman S, Studholme D, Bailey T, Lawson T, et al. Chloroplasts play a central role in plant defence and are targeted by pathogen effectors. Nature Plants. 2015;**1**(6):1-10. DOI: 10.1111/pce.12924

[70] Oliveira TB, Aucique-Pérez CE, Einhardt AM, Rodrigues FÁ. Wheat susceptibility to blast is enhanced by a photosynthetic inhibitor. Journal of Phytopathology. 2021;**169**(10):630-639. DOI: 10.1111/jph.13034

[71] Ajigboye OO, Murchie E, Ray RV. Foliar application of isopyrazam and epoxiconazole improves photosystem II efficiency, biomass and yield in winter wheat. Pesticide Biochemistry and Physiology. 2014;**114**:52-60. DOI: 10.1016/j.pestbp.2014.07.003

[72] Rousseau C, Belin E, Bove E, Rousseau D, Fabre F, Berruyer R, et al. High throughput quantitative phenotyping of plant resistance using chlorophyll fluorescence image analysis. Plant Methods. 2013;**9**(1):17. DOI: 10.1186/1746-4811-9-17

[73] Bürling K, Hunsche M, Noga G. Quantum yield of non-regulated energy dissipation in PSII (Y (NO)) for early detection of leaf rust (*Puccinia triticina*) infection in susceptible and resistant wheat (*Triticum Aestivum* L.) cultivars. Precision Agriculture. 2010;**11**(6):703-716. DOI: 10.1007/s11119-010-9194-1

[74] Kuckenberg J, Tartachnyk I, Schmitz-Eiberger M, Noga G. Early detection of leaf rust and powdery mildew infections on wheat leaves by PAM fluorescence imaging. Precision Agriculture '07. 2007;**8**:515-521. DOI: 10.3920/978-90-8686-603-8

[75] Kim JH, Bhandari SR, Chae SY, Cho MC, Lee JG. Application of maximum quantum yield, a parameter of chlorophyll fluorescence, for early determination of bacterial wilt in tomato seedlings. Horticulture, Environment, and Biotechnology. 2019;**60**:821-829. DOI: 10.1007/s13580-019-00182-0

[76] Grishina A, Sherstneva O, Grinberg M, Zdobnova T, Ageyeva M, Khlopkov A, et al. Pre-symptomatic detection of viral infection in tobacco leaves using pam fluorometry. Plants. 2021;**10**(12):2782. DOI: 10.3390/plants10122782

[77] Moustaka J, Moustakas M. Early-stage detection of biotic and abiotic stress on plants by chlorophyll fluorescence imaging analysis. Biosensors. 2023;**13**(8):796. DOI: 10.3390/bios13080796

[78] Wang L, Poque S, Valkonen JPT. Phenotyping viral infection in sweetpotato using a high-throughput chlorophyll fluorescence and thermal imaging platform. Plant Methods. 2019;**15**:1-14. DOI: 10.1186/s13007-019-0501-1

[79] Aucique-Perez CE, Daza ES, Ávila-Diazgranados RA, Romero HM. Chlorophyll a fluorescence and leaf temperature are early indicators of oil palm diseases. Scientia Agricola. 2019;**77**:e20180106. DOI: 10.1590/1678-992X-2018-0106

[80] Mahlein AK, Alisaac E, Al Masri A, Behmann J, Dehne HW, Oerke EC. Comparison and combination of thermal, fluorescence, and hyperspectral imaging for monitoring fusarium head blight of wheat on spikelet scale. Sensors. 2019;**19**(10):2281. DOI: 10.3390/s19102281

[81] Sun Y, Liu T, Wang X, Hu Y. Chlorophyll fluorescence imaging combined with active oxygen metabolism for classification of similar diseases in cucumber plants. Agronomy. 2023;**13**(3):700. DOI: 10.3390/agronomy13030700

[82] He C, Li X, Liu Y, Yang B, Wu Z, Tan S, et al. Combining multicolor fluorescence imaging with multispectral reflectance imaging for rapid citrus huanglongbing detection based on lightweight convolutional neural network using a handheld device. Computers and Electronics in Agriculture. 2022;**194**:106808. DOI: 10.1016/j.compag.2022.106808

[83] Sapoukhina N, Boureau T, Rousseau D. Plant disease symptom segmentation in chlorophyll fluorescence imaging with a synthetic dataset. Frontiers in Plant Science. 2022;**13**:969205. DOI: 10.3389/fpls.2022.969205

[84] Martínez-Ferri E, Zumaquero A, Ariza MT, Barceló A, Pliego C. Nondestructive detection of white root rot disease in avocado rootstocks by leaf chlorophyll fluorescence. Plant Disease. 2016;**100**(1):49-58. DOI: 10.1094/PDIS-01-15-0062-RE

[85] Liu F, Zhang M, Hu J, Pan M, Shen L, Ye J, et al. Early diagnosis of pine wilt disease in *Pinus thunbergii* based on chlorophyll fluorescence parameters. Forests. 2023;**14**(1):154. DOI: 10.3390/f14010154

[86] Baghbani F, Lotfi R, Moharramnejad S, Bandehagh A, Roostaei M, Rastogi A, et al. Impact of *fusarium verticillioides* on chlorophyll fluorescence parameters of two maize lines. European Journal of Plant Pathology. 2019;**154**:337-346. DOI: 10.1007/s10658-018-01659-x

[87] Weng H, Liu Y, Captoline I, Li X, Ye D, Wu R. Citrus huanglongbing detection based on polyphasic chlorophyll a fluorescence coupled with machine learning and model transfer in two citrus cultivars. Computers and Electronics in Agriculture. 2021;**187**:106289. DOI: 10.1016/j.compag.2021.106289

[88] Petruccelli V, Brasili E, Varone L, Valletta A, Pasqua G. Antifungal activity of dimethyl sulfoxide against *Botrytis cinerea* and phytotoxicity on tomato and lettuce plants. Plant Biosystems - An International Journal Dealing with All Aspects of Plant Biology. 2020;**154**(4):455-462. DOI: 10.1080/11263504.2020.1779846

[89] Suárez JC, Vanegas JI, Contreras AT, Anzola JA, Urban MO, Beebe SE, et al. Chlorophyll fluorescence imaging as a tool for evaluating disease resistance of common bean lines in the western amazon region of Colombia. Plants.

2022;**11**(10):1371. DOI: 10.3390/plants11101371

[90] Warzecha T, Skrzypek E, Adamski T, Surma M, Kaczmarek Z, Sutkowska A. Chlorophyll a fluorescence parameters of hulled and hull-less barley (*Hordeum vulgare* L.) DH lines inoculated with *Fusarium culmorum*. The Plant Pathology Journal. 2019;**35**(2):112. DOI: 10.5423/PPJ.OA.07.2018.0124

Chapter 2

Detection of Phytopathogens in Agricultural Crops Using Nanodiagnostic Techniques

Enespa and Prem Chandra

Abstract

One of the main things restricting yields of crops is diseases that affect plants. Which continue to be the major agricultural threat in the globe and drastically reduce yields of crops internationally, creating serious issues for the availability of food. Despite the fact that chemical-based medication persists as the main tactic for lowering the incidence of agricultural ailments, their frequent usage can make the microorganisms less likely to spread. Consequently, effective screening techniques for the immediate detection of plant-borne pathogens in the initial phases of infection have becoming vital to preserving sustainable farming and adequate nutrition. Quantum dots (QDs), nanoparticles, and nanotechnology have become crucial instruments for the rapid and highly accurate assessment of a specific biochemical marker. Tools including such as biosensors, QDs, nanostructured platforms, nanoimaging, and nanopore DNA sequencing have an opportunity to enhance infection detection's accuracy, precision, and efficiency. They can also make rapid analysis easier and be utilized for crop protection and high-quality monitoring. Additionally, nanodiagnostic tool technology enables professionals to assist producers in avoiding the emergence of pandemics by swiftly and simply identifying potentially hazardous pathogenic organisms in crops.

Keywords: nanotechnology, nanoparticles, agriculture, nano sensor, disease management, quantum dots

1. Introduction

The worldwide transmission of plant pathogens has grown, although the expense associated with identification of pathogens and eradication are relatively low (i.e., less than 3% of the overall expenses associated with agricultural output). According to estimates, 14% of global losses were attributed to pests of insects, 13% to cultivar ailments, and 13% to herbicides [1, 2]. It was estimated that this crop loss costs the United States $2000 billion year. Due to the creation of persistent stressful diseases pathogens limit plant development and output, and precise diagnosis and treatment techniques are costly [3]. Several efforts have been made to produce crops more safely in various conditions using safeguards or optimal managing practices. However, crop protection is essential for ensuring sustained crop output, particularly in challenging situations [4]. In institutions around the world, conventional molecular diagnostic procedures are frequently utilized to detect phyto pathogenic microorganisms having an elevated level of sensitivity as well as specificity. However, almost all of the aforementioned techniques cannot be used in rural areas (on-site detection) or in low-income developing countries [5]. In addition, the use of conventional molecular techniques in emerging economies is constrained by the expensive nature and brief lifespan of particular molecular biology ingredients, such as catalysts and primers [6].

A vast field, the nanotechnology integrates information from sciences such as chemistry, physics, along with other academic areas. Nanotechnology, as defined by Joseph and Morrison [7], is the technique of arranging tiny atoms, molecules, which are or molecular groups into nanostructures to create substances that are capable of carrying out one-of-a-kind or very varied activities. The use of nanotechnologies in farming has the potential to transform farming research as well as bring about the development of innovative instruments for the early and immediate identification of plant-borne infections and illnesses [8]. Because they typically range in size from 1 to 100 nm and can provide enhanced surface-to-volume ratios as well as special chemical, optical, and electrical properties that are not present in their bulk substitutes, nanoparticles are outstanding options for enabling the identification of plant-borne pathogens [9]. Cylindrical fragments, cubes, rods, wires, plates, prisms, core-shell frameworks, and other structures with three dimensions are just a few of the configurations that nanostructures can take [10].

One of the most well-known detecting techniques used by nonmaterial is their capacity to alter their form when exposed to various environmental stimuli, which affects physical and chemical characteristics. It therefore has an enormous capacity to identify infections more quickly and correctly than other pathogen detection techniques now in use [11]. We focused on the recently created uses of nanotechnology instruments for detecting pathogens in plants and diagnosing agricultural illnesses.

2. Types, synthesis, and characterization of nanoparticles used in plants

Microscopic molecules or structures known as nanoparticles usually have a size between 1 and 100 nm. Compared to its mass equivalent, they are more effective, quick to respond, reliable, and operationally engaged [12]. The term "nanotechnology" was first defined scientifically by Norio Taniguchi of Tokyo, Japan. It is the research of atomic or molecular scales of matter. Currently, different kinds of nanoparticles are known to exist [13]. Several factors, including functionalization, surface appearance, chemical nature, physicochemical qualities, scale, genesis and origin, magnetic characteristics, crystalline structure, etc. are taken into account when classifying nanoparticles [14]. Metals and metallic oxide nanoparticles are frequently utilized in crops. Both bottom-up and top-down strategies are two essential ideas that are capable of helping create nanoparticles in physical, chemical, and natural (green synthesis) ways. The first technique uses physical and chemical processes to break down macro-scale or bulky substances into extremely small particles (1–100 nm). Top-down techniques are used to lower the atom size, including mechanical machining, etching, the process of s electro-explosion, and ablation with lasers [15]. The synthesis of green nanoparticles employing living things and plants is a component of the strategy known as bottom-up. Various processes, including chemical vapor deposition, laser pyrolysis, and atomic the process of condensation, are used in this technique for assembling very small atoms into nanoparticles [16]. The physical properties of nanoparticles can be modified by altering reaction circumstances such as chemical concentrations of substances, reaction duration, temperature, and pH. Therefore, anyone is able to create their chosen nanomaterials by altering these parameters [17].

3. Diagnosis of phytopathogens

A method that has promise for locating infections is the advancement or incorporation of molecular detection on a microscopic level. Nanomolecular evaluation, often known as nanodiagnostics, is the application of nanobiotechnology to the diagnosis of plant illnesses [18]. The sequencing of individual DNA molecules especially makes use of a number of tiny devices and nanosystems. Testing methods that examine DNA sequences and identify diseases are getting quicker, easier to modify, and more powerful with the introduction of nano-size devices [19].

It is interesting that this tendency is likely to be supported by novel methods of detection that use nano-biosensors to identify pathogens. Phytopathologists used visual evaluation to recognize plant ailments in the 1980s and 1990s. The standard methods to detect crop infections can take several days, thus researchers require quick detection instruments that can yield results in a few hours [20]. Nanotechnologists and phytopathologists are continually collaborating to create such detecting technologies. Researchers in nanotechnology and/or the phytopathology are working to create a simple, portable assessment that is reliable and does not require a complicated technique so that growers can make use of the transportable laboratory independently to identify particular ailments [21].

Previously created nanoparticles with unique nanoscale properties could represent a significant advance in identifying the presence of contaminants and pathogens [15]. Lab-on-a-chip devices for monitoring micronutrients in water used for irrigation, identifying poisoning in waterways, and regulating the nutritional value of food are also being developed as a result of nanotechnology [22, 23]. The scientific discipline of nano-phytopathology uses nanotechnology in the early phases of disease detection, diagnosis, and management in plants. Nano-phytopathology provides data that can be utilized to identify and track the traits associated with plant illnesses, identify dangerous toxins, evaluate the sustainability of the environment, and spot relationships between pathogenic microbes and their hosts [21, 24]. A more affordable method of protecting plants involves recognizing, detecting, and controlling plant diseases when they are still in their initial phases. Immediate phyto disease detection with nanotechnology-based assays is possible for more cost-effective plant protection control. They are quick and extremely sensitive [25]. Additionally, biological barcoding technique based on nanotechnology provides extremely high specificity for DNA and protein identification. It is possible to create packages for the quick and on-site identification of plant ailments using barcode-based technologies [26]. As a whole, there are many potential uses for nanotechnology with regard to plant monitoring (**Figure 1**).

One of two methods can be used to apply nanoparticles for crop safeguarding: either the microscopic particles themselves protect agricultural products, or they operate as carriers for active ingredients like double-stranded RNA (dsRNA), which can be applied by spraying or wet into the seeds, foliar connective tissue, or roots [27].

Particles as carriers can have a number of benefits, such as (i) an extended shelf life, (ii) improved pesticide aqueous fluctuation, (iii) decreased toxic effects, and (iv) more site-specific incorporation into the pest being targeted [28]. Improved

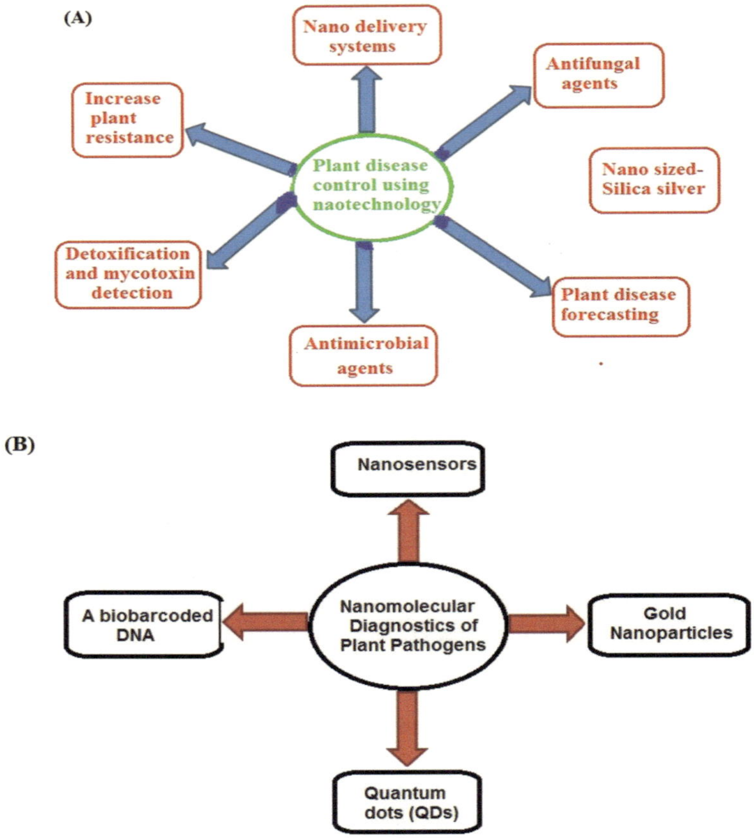

Figure 1.
Potential nanotechnology applications in plant pathology: (A) plant disease control and (B) detection of plant pathogens.

functionality and longevity of small-molecule pesticide under environmental stress (UV and rain) could be another advantage of nanocarriers [29].

This would lead to significantly less usage, minimal negative outcomes, and lower costs (**Figure 2**). This schematic diagram shows different nanomaterials as either protectants or carriers for actives such as insecticides, fungicides, herbicides, or RNA-interference molecules, targeting a wide range of pests and pathogens. It also highlights the potential benefits of nanomaterial applications, such as improved shelf-life, target site-specific uptake, and increased solubility, while decreasing soil leaching and toxicity [27].

4. Diagnostic techniques of phytopathogens

Plant disease detection and identification can be done directly or indirectly. The examination of plant pathogens, such as bacteria, oomycetes, fungi, and viruses, as well as bio molecular markers, such as nucleic acids, proteins, and carbohydrates, isolated from diseased plant tissues, is the norm for direct techniques [30]. Through changes in physiological or histological markers including variations in leaf surface

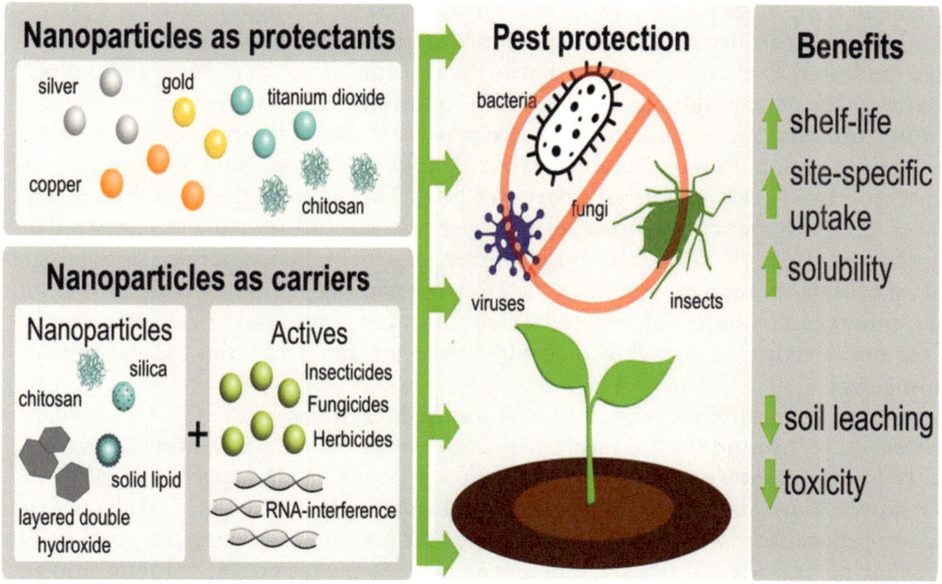

Figure 2.
Nonmaterial as protectants or carriers to provide crop protection.

temperature or humidity, spectroscopic properties of plant tissues, shape, growth rate, and emissions of volatile organic compounds (VOCs), indirect diagnostics can identify plant illnesses [31]. Numerous spectroscopic, electrochemical, or molecular technologies may be used as direct or indirect detection techniques [32].

4.1 Nanobiosensors: a tool for plant pathogens diagnosis

Nano biosensors are nanosensors with bioreceptor devices immobilized that are unique for the molecules that are in the material being studied and have the beneficial properties of being portable, small, and adaptable to real-time screening, accurate, computational, reliable, stable, repeatable, and resiliency for identifying expected and complicated plant health issues [33]. Toxins, viruses, bacteria, fungi, and other bio dangerous substances are some examples of the pollutants that have already been identified and measured using this equipment in the food and agricultural industries. As a result, these nanosensors might have a big impact on precision agricultural methods [34]. Early on-site diagnosis of cultivar diseases using portable nano biosensors can also result from the study of disease epidemiology and the creation of ailment-prevention techniques. For real-time ailments, quality of soil, and crop wellness tracking, these sensors can be positioned throughout the agricultural area and linked to a Global Positioning System [35, 36].

By integrating biotechnological and nanotechnological strategies, biosensors could be created that have greater sensibility, allowing an earlier reaction to changes in the environment and the incidence of diseases [37]. Plant illnesses will be straightforward to manage since we will be able to identify them before any visible symptoms show up. Biosensors will promote precision farming, which will increase productivity and output in agriculture by giving precise data and supporting farmers to make superior choices [38].

Due to their large particular area of surface, size-dependent electrical characteristics, and ability to facilitate machine restructuring, nanostructures have been exploited as creative sensing platforms as a consequence of breakthroughs in nanotechnology and biotechnology [39]. In order to identify microorganisms and mycotoxins, a nano sensing platform was created employing nanomaterials like carbon nanomaterials (nanotubes and graphene), nanofibers, nanotubes, and nanostructured metal oxide nanoparticles [40]. This process speeds up the recognition of microorganisms. Nanostructures based on microfluidic platforms are also effective at promptly and accurately detecting diseases. Food rot triggered by bacterial and fungal pathogens can be detected using a variety of nanostructured devices [41]. Inbaraj and Chen [42], created a structure that contained tens of thousands of nanomaterials that changed color when they got into proximity to food-borne pathogens.

In order to immobilize r-IgG and BSA for the identification of OTA (*Aspergillus ochraceus*), [43] found that an immunoelectrode made of nanoSiO_2 and chitosan worked best. In order to detect very small quantities of AFM1 (up to 0.01 ppb) produced by *A. flavus* in food, [44] used magnetic nanoparticles to build an electrochemical immunosensor [45], created a prototype that measures zearalenone produced by *Fusarium* sp. quickly, cautiously, and precisely by employing a microfluidic chip and an electro kinetic magnetic bead-based electrochemical immunoassay. Furthermore, an ultrasensitive immunosensor developed for the precise and speedy determination of zearalenone in samples of maize silage using continuous-flow technology and multiwall carbon nanotubes.

4.1.1 Colorimetric biosensors

Colorimetric biological sensors are intriguing optical biosensors since they can quickly and accurately identify visually by a color shift the presence of hazardous bacteria in the sample without requiring the application of any extra instruments or chemicals [46]. Similar to how the lateral flow test works, the solution-based colorimetric sensor also measures flow. When target pathogens interact with a colloidal gold nanoparticle-mounted receptor, the color of the nanoparticles changes from red to purple as a result of agglomeration. Horizontal flow immunoassays based on colloidal gold-plated nanoparticles have been developed for a number of agricultural diseases, including Potato Virus X in potatoes, Fusarium species in maize, and *Pantoea stewartii* bacteria in maize [41]. Although it is a serious drawback, all lateral flow assays-based biosensors lacked a high level of sensitivity. Artificial biology and bio sensing technologies are now being used to quickly, accurately, and successfully diagnose plant diseases. However, before they are extensively utilized and applied in plan disease diagnostics, there are a few more aspects and potential obstacles to take into account, such as previous understanding of agricultural systems, a sufficient in-field sampling approach, and a suitable biological concentration of targeted pathogenic [47].

While data-based biological sensors have certainly grown in importance, there are now fewer opportunities for research. An adequate campaign is expected to motivate scientists to refocus their attention on botanical pathogen evaluation and other abnormalities facets employing nanobiosensors due to the importance of plant infectious agent evaluation, the current state of nanobiosensors, and additionally the drawbacks of presently accessible assessment approaches and progressed nanomaterials [34].

4.1.2 Array-based nanosensors

Modern multi-chromophores array-based sensing gatherings, which are analogous to the human sense of smell and have the ability of multiplexed and distinguishing between different investigations [48]. A common variation of the array-based sensor is the "electronic nose" (e-nose), which replaces the chemical sensors in the array with electronic transducers. Because of its capacity to assess chemical fingerprints and cross-reactivity, the array-based technique is useful in discriminating analysis combinations that are similar to one another. It is possible to obtain remarkably precise and prognostic results from supervised or unattended chemometric analysis of the heterogeneous data output of sensing arrays [49]. Unlike e-nose signals, array-based chemical sensors can produce visual readout (such as colorimetric or fluorescent), which is easier to perceive and analyze. Chemical sensor networks are also less expensive, simpler to build, and more immune to outside influences like shifts in temperature and humidity than the bulk of e-nose devices [46]. They also typically have much better chemical specificity because they have their foundation on a particular chemical reaction. Numerous analyteresponsive dyes, such as Lewis acid-based colorants, Bronsted acidic or basic colorants, solvatochromic or vapochromic dyes, and redox indicator colorants, could be embedded in hydrophilic nano porous materials, like altered silica sol-gels, to create chemical detector panels with ease [50].

4.2 Nano barcode assay

Advances in nanotechnology have contributed to the development of a novel technique termed "bio-barcoding" for the identification of enzyme-free ultra-sensitive proteins and DNAs. Protein barcode assays would be more complex, delicate, and thorough than conventional ELISA-based assays that rely on target and sample densities [51]. The nanoparticle-based bio barcode test is sensitive to identifying pathogens and aids in the early detection of plant ailments when compared with different commonly used methods like ELISA, qPCR, etc. Two devices are used in the bio barcode approach [52].

Nanotechnology advancements have led to changes in bio barcode technology. It is an excellent technique for identifying protein complexes and nucleic acids without requiring the use of enzymes or PCR. Modifications have been implemented in a number of categories to expand the usefulness of bio barcodes across multiple domains [53]. DNA barcoding has been suggested as a method for identifying fungus. Standardization, scalability, reproducibility of protein-coding area copies, and accessibility should all be features of a DNA barcode. Gold nanoparticles and the fluorescence bio-barcoding technology are used to identify *P. aeroginosa*. In order to determine a specific DNA at the other side of the probe 1 bio barcode DNA, which acts as a signal identifier, a second probe has been established [25].

As a consequence of this, the two probes formed covalent connections with their appropriate sequences and became triggered. Fluorescence spectroscopy is used to show that this test has a broad linear range of measurement (5–200 ng/mL) and is quick, simple, sensitive, and reliable [54]. When molecular targets use a specific gene as a linker, the oligo-AuNP sensors hybridized with magnetic micro particle (MMP) probes having oligonucleotide functionality. Following magnetic separation of these complicated systems, the oligonucleotides were subsequently freed from the Oligo-AuNP tags. These emitting bio barcodes are quantitatively analyzed by the scientometric test [55]. Under optimal conditions, the test is particularly interesting since it can swiftly detect nucleotides at higher zeptomolar levels and protein complexes at lower up to mole amounts. Similar methods can be established for the fast and on-site detection of agricultural illnesses like viruses, in order to reduce losses in crops [24].

4.3 Quantum dots (QDs)

Quantum dots (QDs), a luminescent semiconductor nanocrystal, produce photons at a certain spectrum depending on their dimension. With magnitude, an infrared light's intensity lengthens [56]. They come with a number of benefits over organic dyes because of their broad stimulation spectra. Quantum dots are resistant to light bleached due to their 10–100 times higher molar attenuation value, distinguishable tunable fluorescence peak, and extended fluorescence lifetime [57]. The aforementioned features of quantum dots make them brighter than conventional fluorophores since they may be activated in a variety of colors from just one source without their emission impulses conflicting [58]. Due to these benefits, QDFRET-based nanosensors have become widely used in the farming and accompanying applications. The measurement of enzyme and nucleic acid activation is the most widespread application for these types of sensors [57]. It was shown that single-cell yeast was able to forming crystallites of cadmium sulfide (CdS) when subjected to cadmium salt exposure, which led to the finding of semiconductor nanomaterials mycosynthesis [59].

Numerous microbes, including fungi, have the potential of biosynthesizing nanoparticles; the internal production of metal nanoparticles may result from the decreased levels of ionized metals by proteins found in the cell membranes. For example, when reacting with a mixture of $CdCl_2$ and $SeCl_4$ at ambient temperature, *Fusarium oxysporum* was capable of to generate extremely luminous CdTe quantum dots [60]. Knudsen et al. [61] showed that QD-based nanosensors may demonstrate a variety of enzyme activity. In recent years, CdTe quantum dots coated with particular antibodies that target the glutathione S transferase (GST) protein found in Polymyxa beta have been employed as biosensors that work [62].

Quantum dots have been placed close enough to one another to make room for the resonant dipole–dipole relationship necessary for fluorescence resonance energy transfer (FRET) due to their shared attraction for antigen and antibody, rhodamine and CdTe [63]. The created immunosensor showed sufficient sensitivity and specificity and was successfully used for rapid screening for plant tests, producing results in less as 30 minutes. In an associated discovery, demonstrated a 100% accuracy for determining the presence of lime plants infested with phytoplasma (*P. aurantifolia*) using a nano biosensor utilizing quantum dots (QDs) [64]. In recent years, a DNA biosensor comprising an artificial quantum dot (QD) and an optical fluorescence resonance energy transfer (FRET) base was created to detect a specific DNA sequence of *Ganoderma boninense*. A single-stranded DNA probe (ssDNA) has been linked to an enhanced quantum dot (5–8 nm) with carboxylic acid groups through an amide coupling [65]. It has been shown that a QD-FRET-based sensor can recognize Polymyxa betae, the sole known vector of the beet necrotic yellow vein virus (BNYVV), which infects beetroot crops. Cadmium selenide (CdSe) quantum dots (QDs) were used [66], as signal amplifiers for more accurate and reliable diagnosis of the banana bunchy top virus. The recombinant coat protein of BBTV was employed to create the primary antibody in the present investigation. This assay's impulses of electricity might be enhanced substantially by CdSe QDs, which makes it suitable to use in a lab [67]. Very little investigation has been performed on applying QD-based biosensors for the detection of crop infections, despite the fact that they have just emerged as an additional kind of sensor and are projected to offer fresh possibilities for accurate identification of pathogenic organisms in plants [68]. It is possible to forecast that quantum dots will play a role in an upcoming breakthrough in the area of agriculture detection of pathogens if the distinctive photo physical properties of quantum dots are put into consideration as an interface element [69].

4.4 Diagnosis of phytopathogens using NPs

Plant illnesses are typically acknowledged directly by human assessors, and then diagnosed through microscopic examination. The pathogen morphology including spore shape, color, and pattern, mycelium, and flowering bodies—is a component of microbial diagnostics. For later treatment and research, pathogens can be separated and grown on specific culture media [70]. These procedures that are regularly carried out in academic institutions and enterprises take time, money, skilled labor, and laboratory space. Their extensive application in identifying illnesses has been constrained by these constraints, particularly in underdeveloped nations. Additionally, genetic, serological, and microbiological investigations are used to find seedling illnesses [71]. For the purpose of tracking plant diseases, a number of direct techniques can be used, including PCR-based nucleic acid identification, fluorescence in situ hybridization, enzyme-linked immunosorbent assay (ELISA), immunofluorescence, flow cytometry, and methods that are indirect like thermography, fluorescence imaging, Hyperspectral approaches, and gas chromatography [25]. Examples of this include diagnosing plant pathogens and investigating fungicide resistance in wheat are both done using molecular techniques. These studies' initial results have aided in the creation of more effective fungicides and resistant hybrids. Additionally, the study of phytopathogen species and how they communicate with plants has been done using molecular approaches [72]. Lateral rotation *Erwinia amylovora* (fire blight), *Phytophthora infestans* (late blight), *Ralstonia solanacearum* (brown rot), Pepino mosaic virus, Tomato mosaic virus, Potato virus Y, and Potato virus X were all identified using ELISA [73]. Techniques that are automated, inexpensive, portable, exact, dependable, and capable of detecting are needed for the advancement of phytopathological disease diagnosis in the years to come. The majority of phytopathological research must be devoted to creating biosensors for the diagnosis of vegetative diseases [74].

The created detectors are capable of detecting variations in tissue of plants color, leaf shape, transpiration rate, canopy morphology, plant density, and fluctuation in wavelength of reflected light from the leaves. They can monitor the reflectance, temperature, and fluorescence of a canopy [75]. There are now being developed new NP-based sensors that could quickly, cheaply, and effectively identify pathogenic organisms in plants. They are giving producers the tools they need to spot infections, volatile substances, chemical residues in crops, and changes in the environment [76]. The most often employed NPs in farming are metal, carbon, and metal oxide NPs. The widespread use of nanosensing tools coupled with fresh, cutting-edge approaches will undoubtedly revolutionize farming. The following describes a few different kinds of nanosensors [77].

4.5 Diagnosis through magnetic NPs

Due to their size being similar to the proportions of the magnetically domain, magnetic NPs have unique features. They exhibit both single-domain ferromagnetism and superparamagnetism as dual behaviors. Although it is not a brand-new idea, pathology of plants has not entirely investigated the use of magnetic nanoparticles (NPs) in biological research [78].

It has been possible to see the route, accumulation, and transportation of magnetic nanoparticles inside plant cells by using carbon-coated magnetic NPs. Using magnetic nanotags on a spin valve sensor surface immobilized with trap antibodies, a unique NP immunoassay was created to detect the quantity of mycotoxins, which in vegetation in instantaneously [79]. Super paramagnetic NPs were used to create a dependable and quick ELISA approach that decreased the time needed for coating, enzyme preventing, and competitiveness. Magnetic NPs that have been packed with vegetation-protection compounds can be used in a novel way to fill the within-cell gaps of plant cells [80]. Such particles can be directed to disease-specific locations in crops with the aid of strong magnets. They can be utilized to create developed, target-specific compounds for discharge and to create a method to monitor the motion of internalized magnetic NPs. For instance, carbon-coated magnetic NPs have been created to enable visualization of the transportation channel and NP deposit within the plant [81].

4.6 Diagnosis through polymeric NPs

The phrase "polymer nanoparticle" especially encompasses tiny spheres and nanocapsules made from either synthetic or natural polymers, including as albumin, alginates, chitosan, DNA, gelatin, gliadin, and poly (L-lactides). Synthetic polymers include polyacrylamide, polyacrylate, polyanhydrides, and polycaprolactone [82]. While conductive substances like polyacetylene, polyaniline, or polypyrrole transform biological messages into electrical signals, polymeric nanomaterials are especially utilized for creating biosensors. Conjugated polymer nanoparticles have been shown to be effective as a transferring carrier for the transport of short interfering RNA (siRNA) in plant protoplasts. Other transfected techniques on the market result in a 40% loss of viable protoplast [83, 84]. However, during the first 24 hours after their delivery, CPN-assisted delivery only results in a 5–25% loss of protoplasts. Due to its biological compatibility, high surface area, and significant porosity, findings support the use of chitosan nanofibrous membranes in the immobilization of enzymes [85].

The technique center around chitosan is now able to be utilized to create biosensors. By using a carbon electrode transformed with a tyrosinase-Fe_3O_4 magnetic nanoparticles-chitosan nanobiocomposite film, phenolic chemicals were detected using a tyrosinase biosensor [86] Glutathione has a widely recognized function as a signaling molecule in plant's defense mechanism. By permanently immobilizing glutathione oxidase on the surface of gold-coated magnetic nanoparticles-modified Pt electrodes, amperometric nanobiosensors have been produced to determine glutathione content. The ability to identify the presence of glutathione was highly sensitive when glutathione oxidase, chitosan, and gold-coated magnetic nanoparticles were integrated [87].

4.7 Diagnosis through carbon NPs

Among the more prevalent substances on Earth, carbon serves as the building block for the majority of biological processes. Compared to other elements, carbon has several unique benefits. As a consequence, of all of the synthesized nanomaterials, carbon NPs are the most commonly employed NPs [9]. Carbon allotropes known as CNTs are cylindrical and promote the development of plants by improving water intake. CNTs are employed as parts in biosensors because of their distinct electrochemical characteristics. Their characteristics comprise having the capacity to combine using any kind of chemical organisms, a higher length to dimension ratio, and fast electron transfer kinetics [88, 89]. Since most phytopathological disorders can be connected to alterations in the absorption and utilization of fragrant chemicals, CNTs are being used to detect metabolites from plants and diagnose plant diseases. Phenolic chemicals, which are volatile compounds, are crucial in controlling diseases of plants [90].

Because phenols are easily oxidized, they can be identified with amperometric and potentiometric instruments. However, electrode corrosion in electrochemical sensors is a significant downside brought on by phenol being exposed, which produces dimeric or polymeric oxidation products [91]. By collecting biomolecules like nucleic acids, electrodes built of CNT minimize the effect of the electrode's outer contamination. In order to detect phenolic compounds, CNTs are either utilized alone or in conjunction with enzyme. The main hormone controlling the development of plants, indole-3-acetic acid (IAA), is linked to numerous processes related to development. The coating of DHP (dihydropyran)-stabilized MWCNT scattering onto a carbon electrode has been used to create sensors for IAA identification. CNTs have been used to construct a variety of electrolytic nanosensors, including DNA hybridization biological sensors and amperometric enzymatic electrodes [92].

5. Metal or metalloid nanoparticles

Inorganic material nanostructures including Ag, Au, Si, and other metal oxide nanoparticles belong to a class of nanosensors that are able to identify and measure molecular targets of interest using signal transducers and identification ligands like antibodies or DNA oligos [93]. These tiny particles can be used as special sensing instruments to examine relationships in vitro or in living systems of plants between the nanoparticles and bioanalytes that are specifically relevant for agricultural diseases. Changes in the surface-enhanced optical properties, which result from the localized surfaces plasmon resonance (LSPR) of tiny particles, are frequently used to record distinctive sensor responses [94]. To create accurate and target-specific nanobiosensors, a wide range of oligonucleotide- or protein-based formed themselves nanomaterials have been investigated [95]. For instance, substrate enhanced Raman spectroscopy (SERS) has been frequently employed to identify different plant illnesses or generated poisons utilizing Ag nanorods as sensor substrates. Matrix-assisted laser desorption/ionization mass spectrometry (MALDI MS) was enhanced significantly to identify plant-associated bacteria in soil and carrots by the incorporation of Pt nanosensors personalized with IgG antibodies [96]. It has been established that fluorescent silica nanoparticles coupled with a secondary antibody and doped with a Ru (II) complex may effectively identify plant-borne pathogens like *X. campestris* that cause bacterial spot sickness in nightshades plants. Although the procedure was extremely delicate, the concerns about the environment raised by the usage of potentially hazardous heavy metal complexes were not taken into consideration [97].

The genomic DNA of *R. solanacearum*, a transmitted by soil organism that could trigger potato wilt due to bacteria, was detected using au nanoparticles modified with a particular single-stranded DNA at concentrations as low as 15 ng. However,

Sensor component	Sensor fabrication	Target	Specificity	Detection mechanism	Toxicity	Reference
Au NP[a]-ssDNA	Nanoparticle functionalization and DNA hybridization	*P. syringae*	Unresponsive to *B. cinerea* and *F. oxysporum*	Electrochemistry	Low	[100]
Ag NP-ssDNA	Nanoparticle functionalization and DNA hybridization	*P. ramorum* and *P. lateralis*	Unresponsive to *P. lateralis*	SERS[b]	Low	[101]
Ag NR[c]	Surface functionalization	Aflatoxins	Discriminable among Aflatoxin B1, B1, G1 and G2	SERS	Low	[102]
Pt NP-IgG[d]	Surface immunological functionalization	*B. thuringiensis* and *B. subtilis*	Discriminable between two bacteria	MALDI-TOF MS[e]	Low	[103–105]
Si NP-Rubpy-IgG	Surface immunological functionalization	*X. campestris*	NA	Fluorescence quenching	Moderate	[31]
Au NP-ssDNA	Surface functionalization	*R. solanacearum*	No specific band revealed by PCR	Colorimetric	Low	[1]
CdSe/ZnS-MPA[f] QD[g]	Surface functionalization	F. oxysporum	NA	Fluorescence	High	[105]
CdSe-PEIh QD	Surface functionalization	*Arabidopsis thaliana*	NA	Fluorescence	High	[105]
CD[i]	Surface functionalization	*V. mali*	No interference from ions	Fluorescence	Low	[106]
CD	Surface functionalization	*F. avenaceum*	Discriminable *between P. aeruginosa* and *F. avenaceum*	Fluorescence	Low	[31]
CdTe QD-CD	Surface immunological functionalization	*Citrus tristeza*	NA	FRET[j]	High	[107]
CdTe QD-Rd[k]	Surface immunological functionalization	*Citrus tristeza*	NA	FRET	High	[108]

a Nanoparticle, b Surface-enhanced Raman spectroscopy, c Tris(bipyridine)ruthenium(II)chloride, d Immunoglobulin G, e Matrix-assisted laser desorption/ionization-time of flight mass spectroscopy, f 3-Mercaptopropionic acid, g Quantum dot, h Polyethylenimine, i Carbon dot, j Förster resonance energy transfer, k Rhodamine 123. l Single-walled carbon nanotube, m Volatile organic compounds.

Table 1.
Summary of nanoparticle-based sensors for plant pathogen or disease marker detection.

additional quantitative calibration information at the lower observable spectrum was not accessible, despite the test exhibiting an adequate degree of sensitivity toward the target DNA [98]. Additionally, pesticides, fertilizers, and numerous additional crop-related characteristics are detected and controlled by nanosensors or nanobiosensors, which provides current data for precise choice-making and agricultural management. For sensor use in agricultural precision farming, extra field validation tests are nonetheless required [99].

A list of representative nanosensors or biosensors used for plant disease diagnosis, their detection mechanisms, and associated performance is summarized in **Table 1**.

6. Post-harvest plant products and food

Films produced using cardamom or oregano oil, nanoparticles of zinc or calcium, or additional substances that destroy germs are currently being tested as antibacterial packaging for consumable foods [109]. Another effort to ensure the integrity of food is environmentally friendly packaging, which uses nanofibers produced from organic maize or lobster shells (both of which are antibacterial and compostable). Packing materials with endurance, protective qualities, and resilience to both heat as well as cold are necessary for enhanced food packing [110]. These are achieved with the use of nanocomposite substances.

The 'Durethan' nanocomposite coating, created by Bayer Polymers, is loaded with silicate nanoparticles to inhibit the entry into oxygen and other pollutants while maintaining humidity and avoiding food spoilage [111]. In future generations, contaminating will be prevented by adding silver, magnesium oxide, or zinc oxide nanoparticles (which can destroy hazardous microbes) to foodstuff or beverage packaging. Prospects for the future call for the ability to incorporate nano-sensors to food packaging to impart antimicrobial properties [112]. These tiny sensors may be utilized to identify pollutants, infections, and poisons in nourishment, according to McHugh. Upcoming packaging for food may contain RFID tags for radio frequency identification (RFID). These allow the identification of hundreds of tags in a second and do not require the line of sight processing like bar codes do. The properties of food packaging are currently being enhanced by the incorporation of nanowheels, nanofibers, and nanotubes [113].

7. Conclusions

Since only a few years ago, the number of people has been growing at a pace that is unprecedented, leading to the growth of industry, the reduction of land used for agriculture, and an increasing urbanization of territory. Immediate and highly sensitive disease probing made possible by nanotechnology technologies provide substantial advancements in the field of agricultural detection of pathogenic organisms. The present-day plant technology for diagnosis continues to face certain substantial obstacles, though. Whenever any tiny sensors are utilized in their natural environment, safety risks have to be addressed since some micron-sized particles like QDs may be hazardous. Health and safety evaluation and monitoring must be more stringent. The starting forecast of longer-lasting and more resilient sensors that can tolerate an extensive variety of ambient fluctuations is necessary for nano-based instruments. This calls for more thorough research on the novel sensor ingredients, such as nonmaterial and substances resilient to the surroundings.

Author details

Enespa[1] and Prem Chandra[2*]

1 Department Plant Pathology, MPDC, University of Lucknow, Lucknow, U.P., India

2 School of Agriculture, Indira Gandhi National Open University, New Delhi, India

*Address all correspondence to: p.chandrabbau@gmail.com

References

[1] Shivashakarappa K, Reddy V, Tupakula VK, Farnian A, Vuppula A, Gunnaiah R. Nanotechnology for the detection of plant pathogens. Plant Nano Biology. 2022;**2**:100018

[2] Lahlali R, Ezrari S, Radouane N, Kenfaoui J, Esmaeel Q, El Hamss H, et al. Biological control of plant pathogens: A global perspective. Microorganisms. 2022;**10**(3):596

[3] Buja I, Sabella E, Monteduro AG, Chiriacò MS, De Bellis L, Luvisi A, et al. Advances in plant disease detection and monitoring: From traditional assays to in-field diagnostics. Sensors. 2021;**21**(6):2129

[4] Deguine JP, Aubertot JN, Flor RJ, Lescourret F, Wyckhuys KA, Ratnadass A. Integrated pest management: Good intentions, hard realities. A review. Agronomy for Sustainable Development. 2021;**41**(3):38

[5] Hariharan G, Prasannath K. Recent advances in molecular diagnostics of fungal plant pathogens: A mini review. Frontiers in Cellular and Infection Microbiology. 2021;**10**:600234

[6] Mondini L, Noorani A, Pagnotta MA. Assessing plant genetic diversity by molecular tools. Diversity. 2009;**1**(1):19-35

[7] Joseph T, Morrison M. Nanotechnology in Agriculture and Food. A Nanoforum Report. Institute of Nanotechnology; 2006. Available from: https://www.nanoforum.org

[8] Alaa YG, Al TM. Applications of Nanotechnology in Agriculture. United Kingdom: Intech Open; 2020. pp. 1-14

[9] Khan Y, Sadia H, Ali Shah SZ, Khan MN, Shah AA, Ullah N, et al. Classification, synthetic, and characterization approaches to nanoparticles, and their applications in various fields of nanotechnology: A review. Catalysts. 2022;**12**(11):1386

[10] Arenal R, Henrard L, Roiban L, Ersen O, Burgin J, Treguer-Delapierre M. Local plasmonic studies on individual core–shell gold–silver and pure gold nano-bipyramids. The Journal of Physical Chemistry. 2014;**118**(44):25643-25650

[11] Ravichandran R. Nanotechnology applications in food and food processing: Innovative green approaches, opportunities and uncertainties for global market. International Journal of Green Nanotechnology: Physics and Chemistry. 2010;**1**(2):72-96

[12] Altammar KA. A review on nanoparticles: Characteristics, synthesis, applications, and challenges. Frontiers in Microbiology. 2023;**14**:1155622

[13] Mulvaney P. Nanoscience vs nanotechnology-defining the field. ACS Nano. 2015;**9**(3):2215-2217

[14] Joudeh N, Linke D. Nanoparticle classification, physicochemical properties, characterization, and applications: A comprehensive review for biologists. Journal of Nanobiotechnology. 2022;**20**(1):262

[15] Baig N, Kammakakam I, Falath W. Nanomaterials: A review of synthesis methods, properties, recent progress, and challenges. Materials Advances. 2021;**2**(6):1821-1871

[16] Jassal PS, Kaur D, Prasad R, Singh J. Green synthesis of titanium dioxide

nanoparticles: Development and applications. Journal of Agriculture and Food Research. 2022;**10**:100361

[17] Khan I, Saeed K, Khan I. Nanoparticles: Properties, applications and toxicities. Arabian Journal of Chemistry. 2019;**12**(7):908-931

[18] Govindaraju M, Li Y, Zhang M. Emerging bacterial disease (leaf scald) of sugarcane in China: Pathogenesis, diagnosis, and management. In: Plant Diseases-Current Threats and Management Trends. London, UK: IntechOpen; 2019. pp. 117-128

[19] Jain KK. Nanomolecular diagnostics. In: The Handbook of Nanomedicine. York, NY: Humana Press; 2017

[20] Dyussembayev K, Sambasivam P, Bar I, Brownlie JC, Shiddiky MJ, Ford R. Biosensor technologies for early detection and quantification of plant pathogens. Frontiers in Chemistry. 2021;**9**:636245

[21] Das K, Jhan PK, Das SC, Aminuzzaman FM, Ayim BY. Nanotechnology: Past, present and future prospects in crop protection. In: Technology in Agriculture. 2021. pp. 1-22

[22] Yadav A, Yadav K, Ahmad R, Abd-Elsalam KA. Emerging Frontiers in nanotechnology for precision agriculture: Advancements, hurdles and prospects. Agrochemicals. 2023;**2**(2):220-256

[23] Yoon JY, Kim B. Lab-on-a-chip pathogen sensors for food safety. Sensors (Basel). 2012;**12**(8):10713-10741

[24] Ray MK, Mishra AK, Mohanta YK, Mahanta S, Chakrabartty I, Kungwani NA, et al. Nanotechnology as a promising tool against phytopathogens: A futuristic approach to agriculture. Agriculture. 2023;**13**(9):1856

[25] Fang Y, Ramasamy RP. Current and prospective methods for plant disease detection. Biosensors. 2015;**5**(3):537-561

[26] Munir S, Ahmed S, Ibrahim M, Khalid M, Ojha SC. A spellbinding interplay between biological barcoding and nanotechnology. Frontiers in Bioengineering and Biotechnology. 2020;**8**:883

[27] Worrall EA, Hamid A, Mody KT, Mitter N, Pappu HR. Nanotechnology for plant disease management. Agronomy. 2018;**8**(12):285

[28] Damalas CA, Eleftherohorinos IG. Pesticide exposure, safety issues, and risk assessment indicators. International Journal of Environmental Research and Public Health. 2011;**8**(5):1402-1419

[29] Asif M, Islam S, Malik MA, Dar ZM, Masood A, Shafi S, et al. Nano pesticides application in agriculture and their impact on environment. In: Agricultural Development in Asia-Potential Use of Nano-Materials and Nano-Technology. London, UK: Intech Open; 2021. pp. 43-53

[30] Marwal A, Gaur RK. Nanophytovirology: An Emerging Field for Disease Management. London, UK: IntechOpen; 2020. pp. 19-29

[31] Li Z, Yu T, Paul R, Fan J, Yang Y, Wei Q. Agricultural nanodiagnostics for plant diseases: Recent advances and challenges. Nanoscale Advances. 2020;**2**(8):3083-3094

[32] Boelens HF, Dijkstra RJ, Eilers PH, Fitzpatrick F, Westerhuis JA. New background correction method for liquid chromatography with diode array detection, infrared spectroscopic detection and Raman spectroscopic detection. Journal of Chromatography A. 2004;**1057**(1-2):21-30

[33] Pérez-López B, Merkoçi A. Nanomaterials based biosensors for food analysis applications. Trends in Food Science & Technology. 2011;**22**(11):625-639

[34] Mathivanan S. Perspectives of nanomaterials and nanobiosensors in food safety and agriculture. UK: Intech Open; 2021. pp. 197-217

[35] Ngashangva L, Hemdan BA, El-Liethy MA, Bachu V, Minteer SD, Goswami P. Emerging bio analytical devices and platforms for rapid detection of pathogens in environmental samples. Micromachines. 2022;**13**(7):1083

[36] Yusof NA, Isha A. Nanosensors for early detection of plant diseases. In: Nanomaterials for Agriculture and Forestry Applications. UK: Elsevier; 2020. pp. 407-419

[37] Ramesh M, Janani R, Deepa C, Rajeshkumar L. Nanotechnology-enabled biosensors: A review of fundamentals, design principles, materials, and applications. Biosensors. 2022;**13**(1):40

[38] Fang W, Wu J, Cheng M, Zhu X, Du M, Chen C, et al. Diagnosis of invasive fungal infections: Challenges and recent developments. Journal of Biomedical Science. 2023;**30**(1):42

[39] Díez-Pascual AM. Surface engineering of nanomaterials with polymers, biomolecules, and small ligands for nano medicine. Materials. 2022;**15**(9):3251

[40] Mohammadi Z, Jafari SM. Detection of food spoilage and adulteration by novel nanomaterial-based sensors. Advances in Colloid and Interface Science. 2020;**286**:102297

[41] Enespa, Dwivedi SK. Effectiveness of some antagonistic fungi and botanicals against *Fusarium solani* and *Fusarium oxysporum* f. sp. lycopersici infecting brinjal and tomato plants. Asian Journal of Plant Pathology. 2014;**8**(1):18-25

[42] Inbaraj BS, Chen BH. Nanomaterial-based sensors for detection of foodborne bacterial pathogens and toxins as well as pork adulteration in meat products. Journal of Food and Drug Analysis. 2016;**24**(1):15-28

[43] Kaushik A, Solanki PR, Ansari AA, Sumana G, Ahmad S, Malhotra BD. Iron oxide-chitosan nanobiocomposite for urea sensor. Sensors and Actuators B: Chemical. 2009;**138**(2):572-580

[44] Paniel N, Radoi A, Marty JL. Development of an electrochemical biosensor for the detection of aflatoxins M1 in milk. Sensors. 2010;**10**(10):9439-9448

[45] Hervas-Stubbs S, Perez-Gracia JL, Rouzaut A, Sanmamed MF, Le Bon A, Melero I. Direct effects of type I interferon on cells of the immune system. Clinical Cancer Research. 2011;**17**(9):2619-2627

[46] Wu Y, Feng J, Hu G, Zhang E, Yu HH. Colorimetric sensors for chemical and biological sensing applications. Sensors. 2023;**23**(5):2749

[47] Lowdon JW, Diliën H, Singla P, Peeters M, Cleij TJ, van Grinsven B, et al. MIPs for commercial application in low-cost sensors and assays–An overview of the current status quo. Sensors and Actuators B: Chemical. 2020;**325**:128973

[48] Geng Y, Peveler WJ, Rotello VM. Array-based "chemical nose" sensing in diagnostics and drug discovery. Angewandte Chemie International Edition. 2019;**58**(16):5190-5200

[49] Wilson AD, Baietto M. Applications and advances in

electronic-nose technologies. Sensors. 2009;**9**(7):5099-5148

[50] Suslick KS, Rakow NA, Kosal ME, McNamara WB III, Sen A. Chemsensing: A colorimetric array detector. In: Proceeding International Society. Olfaction Electron. Noses. 2003;**6**:46-52

[51] Wang Y, Jin M, Chen G, Cui X, Zhang Y, Li M, et al. Bio-barcode detection technology and its research applications: A review. Journal of Advanced Research. 2019;**20**:23-32

[52] Venbrux M, Crauwels S, Rediers H. Current and emerging trends in techniques for plant pathogen detection. Frontiers in Plant Science. 2023;**14**:1120968

[53] Abu-Salah KM, Zourob MM, Mouffouk F, Alrokayan SA, Alaamery MA, Ansari AA. DNA-based nanobiosensors as an emerging platform for detection of disease. Sensors. 2015;**15**(6):14539-14568

[54] Arai MS, de Camargo AS. Exploring the use of up conversion nanoparticles in chemical and biological sensors: From surface modifications to point-of-care devices. Nanoscale Advances. 2021;**3**(18):5135-5165

[55] Rajasundari K, Ilamurugu K. Nanotechnology and its applications in medical diagnosis. Journal of Basic and Applied Chemistry. 2011;**1**(2):26-32

[56] Pietryga JM, Park YS, Lim J, Fidler AF, Bae WK, Brovelli S, et al. Spectroscopic and device aspects of nanocrystal quantum dots. Chemical Reviews. 2016;**116**(18):10513-10622

[57] Stanisavljevic M, Krizkova S, Vaculovicova M, Kizek R, Adam V. Quantum dots-fluorescence resonance energy transfer-based nanosensors and their application. Biosensors and Bioelectronics. 2015;**74**:562-574

[58] Brkić S. Applicability of quantum dots in biomedical science. In: Ionizing Radiation Effects and Applications. New York: Humana Press; 2018

[59] Ghasempour A, Dehghan H, Ataee M, Chen B, Zhao Z, Sedighi M, et al. Cadmium sulfide nanoparticles: Preparation, characterization, and biomedical applications. Molecules. 2023;**28**(9):3857

[60] Dwivedi SK, Enespa. In vitro efficacy of some fungal antagonists against *Fusarium solani* and *Fusarium oxysporum* f. sp. lycopersici causing brinjal and tomato wilt. International Journal of Biological & Pharmaceutical Research. 2013;**4**(1):46-52

[61] Knudsen BR, Jepsen ML, Ho YP. Quantum dot-based nanosensors for diagnosis via enzyme activity measurement. Expert Review of Molecular Diagnostics. 2013;**13**(4):367-375

[62] Safarpour H, Safarnejad MR, Tabatabaei M, Mohsenifar A, Rad F, Basirat M, et al. Development of a quantum dots FRET-based biosensor for efficient detection of Polymyxa betae. Canadian Journal of Plant Pathology. 2012;**34**(4):507-515

[63] Shamirian A, Ghai A, Snee PT. QD-based FRET probes at a glance. Sensors. 2015;**15**(6):13028-13051

[64] Rad F, Mohsenifar A, Tabatabaei M, Safarnejad MR, Shahryari F, Safarpour H, et al. Detection of Candidatus *Phytoplasma aurantifolia* with a quantum dots fret-based biosensor. Journal of Plant Pathology. 2012;**94**(3):525-534

[65] Mohd Bakhori N, Yusof NA, Abdullah AH, Hussein MZ. Development of a fluorescence resonance energy transfer (FRET)-based DNA biosensor for detection of synthetic oligonucleotide of *Ganoderma boninense*. Biosensors. 2013;**3**(4):419-428

[66] Majumder S, Bhattacharya B, Singh PK, Johari S, Singh B, Rahman R. Impedimetric detection of banana bunchy top virus using CdSe quantum dots for signal amplification. SN Applied Sciences. 2020;**2**:1-7

[67] Thomas JE, Dietzgen RG. Purification, characterization and serological detection of virus-like particles associated with banana bunchy top disease in Australia. Journal of General Virology. 1991;**72**(2):217-224

[68] Hong S, Lee C. The current status and future outlook of quantum dot-based biosensors for plant virus detection. The Plant Pathology Journal. 2018;**34**(2):85

[69] James Singh K, Ahmed T, Gautam P, Sadhu AS, Lien DH, Chen SC, et al. Recent advances in two-dimensional quantum dots and their applications. Nanomaterials. 2021;**11**(6):1549

[70] Henson JM, French R. The polymerase chain reaction and plant disease diagnosis. Annual Review of Phytopathology. 1993;**31**(1):81-109

[71] Vasala A, Hytönen VP, Laitinen OH. Modern tools for rapid diagnostics of antimicrobial resistance. Frontiers in Cellular and Infection Microbiology. 2020;**10**:308

[72] Hahn M. The rising threat of fungicide resistance in plant pathogenic fungi: Botrytis as a case study. Journal of Chemical Biology. 2014;**7**:133-141

[73] Whisson SC, Boevink PC, Wang S, Birch PR. The cell biology of late blight disease. Current Opinion in Microbiology. 2016;**34**:127-135

[74] Srinivasan B, Tung S. Development and applications of portable biosensors. Journal of Laboratory Automation. 2015;**20**(4):365-389

[75] West JS, Bravo C, Oberti R, Moshou D, Ramon H, McCartney HA. Detection of fungal diseases optically and pathogen inoculum by air sampling. In: Oerke EC, Gerhards R, Menz G, Sikora R, editors. Precision Crop Protection - The Challenge and Use of Heterogeneity. Dordrecht: Springer; 2010. pp. 135-149

[76] Javaid M, Haleem A, Singh RP, Rab S, Suman R. Exploring the potential of nanosensors: A brief overview. Sensors International. 2021;**2**:100130

[77] Ijaz M, Khan F, Ahmed T, Noman M, Zulfiqar F, Rizwan M, et al. Nanobiotechnology to advance stress resilience in plants: Current opportunities and challenges. Materials Today Bio. 2023;**22**:100759

[78] Wallyn J, Anton N, Vandamme TF. Synthesis, principles, and properties of magnetite nanoparticles for in vivo imaging applications—A review. Pharmaceutics. 2019;**11**(11):601

[79] Cifuentes Z, Custardoy L, de la Fuente JM, Marquina C, Ibarra MR, Rubiales D, et al. Absorption and translocation to the aerial part of magnetic carbon-coated nanoparticles through the root of different crop plants. Journal of Nanobiotechnology. 2010;**8**:1-8

[80] Bilal M, Iqbal HM, Adil SF, Shaik MR, Abdelgawad A, Hatshan MR, et al. Surface-coated magnetic nanostructured

materials for robust bio-catalysis and biomedical applications-A review. Journal of Advanced Research. 2022;**38**:157-177

[81] Naidu S, Singh IK, Singh A. Microbial synthesis of magnetic nanoparticles for plant science and agriculture. Plant Nano Biology. 2023;**4**:100036

[82] Alkubaisi NAO, Aref NMA. The intervention of gold nanoparticles (AuNPs) interactions lead to the disappearing of virus particles. In: Atlas of Ultrastructure Interaction Proteome between Barley Yellow Dwarf Virus and Gold Nanoparticles. London, UK: IntechOpen; 2021. pp. 97-443

[83] El-Said WA, Abdelshakour M, Choi JH, Choi JW. Application of conducting polymer nanostructures to electrochemical biosensors. Molecules. 2020;**25**(2):307

[84] Kuznetsova LS, Arlyapov VA, Plekhanova YV, Tarasov SE, Kharkova AS, Saverina EA, et al. Conductive polymers and their nanocomposites: Application features in biosensors and biofuel cells. Polymers. 2023;**15**(18):3783

[85] Agnihotri SA, Mallikarjuna NN, Aminabhavi TM. Recent advances on chitosan-based micro-and nanoparticles in drug delivery. Journal of Controlled Release. 2004;**100**(1):5-28

[86] Wang S, Tan Y, Zhao D, Liu G. Amperometric tyrosinase biosensor based on Fe_3O_4 nanoparticles–chitosan nanocomposite. Biosensors and Bioelectronics. 2008;**23**(12):1781-1787

[87] Hasanuzzaman M, Nahar K, Anee TI, Fujita M. Glutathione in plants: Biosynthesis and physiological role in environmental stress tolerance. Physiology and Molecular Biology of Plants. 2017;**23**:249-268

[88] Ferrier DC, Honeychurch KC. Carbon nanotube (CNT)-based biosensors. Biosensors. 2021;**11**(12):486

[89] Sharma R, Dewanjee S, Kole C. Utilization of nanoparticles for plant protection. In: Kole C, Kumar D, Khodakovskaya M, editors. Plant Nanotechnology. Cham: Springer; 2016

[90] Ziyatdinova G, Guss E, Yakupova E. Electrochemical sensors based on the electropolymerized natural phenolic antioxidants and their analytical application. Sensors. 2021;**21**(24):8385

[91] Zhao Q, Gan Z, Zhuang Q. Electrochemical sensors based on carbon nanotubes. Electroanalysis: An International Journal Devoted to Fundamental and Practical Aspects of Electroanalysis. 2002;**14**(23):1609-1613

[92] Leveau JH, Lindow SE. Utilization of the plant hormone indole-3-acetic acid for growth by *Pseudomonas putida* strain 1290. Applied and Environmental Microbiology. 2005;**71**(5):2365-2371

[93] Malik P, Gupta R, Malik V, Ameta RK. Emerging nanomaterials for improved biosensing. Measurement: Sensors. 2021;**16**:100050

[94] Unser S, Bruzas I, He J, Sagle L. Localized surface plasmon resonance biosensing: Current challenges and approaches. Sensors. 2015;**15**(7):15684-15716

[95] Huang X, Zhu Y, Kianfar E. Nano biosensors: Properties, applications and electrochemical techniques. Journal of Materials Research and Technology. 2021;**12**:1649-1672

[96] Prabakaran E, Pillay K. Nanomaterials for latent fingerprint

detection: A review. Journal of Materials Research and Technology. 2021;**12**:1856-1885

[97] Gubala V, Giovannini G, Kunc F, Monopoli MP, Moore CJ. Dye-doped silica nanoparticles: Synthesis, surface chemistry and bio applications. Cancer Nanotechnology. 2020;**11**(1):1-43

[98] Ferreira V, González M, Pianzzola MJ, Coll NS, Siri MI, Valls M. Molecular detection of Ralstonia solanacearum to facilitate breeding for resistance to bacterial wilt in potato. In: Dobnik D, Gruden K, Ramšak Ž, Coll A, editors. *Solanum tuberosum*. Methods in Molecular Biology. Vol. 2354. New York, NY: Humana; 2021

[99] Duhan JS, Kumar R, Kumar N, Kaur P, Nehra K, Duhan S. Nanotechnology: The new perspective in precision agriculture. Biotechnology Reports. 2017;**15**:11-23

[100] Ma X, Li X, Luo G, Jiao J. DNA-functionalized gold nanoparticles: Modification, characterization, and biomedical applications. Frontiers in Chemistry. 2022;**10**:1095488

[101] Fahrina A, Arahman N, Aprilia S, Bilad MR, Silmina S, Sari WP, et al. Functionalization of PEG-AgNPs hybrid material to alleviate biofouling tendency of polyethersulfone membrane. Polymers. 2022;**14**(9):1908

[102] Lin B, Kannan P, Qiu B, Lin Z, Guo L. On-spot surface enhanced Raman scattering detection of Aflatoxin B1 in peanut extracts using gold nanobipyramids evenly trapped into the AAO nanoholes. Food Chemistry. 2020;**307**:125528

[103] Chen M, Wei X, Zhang J, Zhou H, Chen N, Wang J, et al. Differentiation of *Bacillus cereus* and *Bacillus thuringiensis* using genome-guided MALDI-TOF MS based on variations in ribosomal proteins. Microorganisms. 2022;**10**(5):918

[104] Wang H, Song D, Zhou Y, Liu J, Zhu A, Long F. Fluorescence enhancement of CdSe/ZnS quantum dots induced by mercury ions and its applications to the on-site sensitive detection of mercury ions. Microchimica Acta. 2021;**188**(6):215

[105] Pavlicek A, Neubauer S, Zafiu C, Huber-Humer M, Ehmoser EK, Part F. The use and detection of quantum dots as nanotracers in environmental fate studies of engineered nanoparticles. Environmental Pollution. 2023;**317**:120461

[106] Wang Y, Darling SB, Chen J. Selectivity of per-and polyfluoroalkyl substance sensors and sorbents in water. ACS Applied Materials & Interfaces. 2021;**13**(51):60789-60814

[107] Li G, Fei X, Liu H, Gao J, Nie J, Wang Y, et al. Fluorescence and optical activity of chiral CdTe quantum dots in their interaction with amino acids. ACS Nano. 2020;**14**(4):4196-4205

[108] Gao W, Wang X, He Y, Yu H, Zheng Y, Yin R, et al. Sub-ppm NO_2 gas sensing in CdTe quantum dots functionalized hollow-core anti-resonant fiber. Sensors and Actuators B: Chemical. 2024;**405**:135350

[109] Duda-Chodak A, Tarko T, Petka-Poniatowska K. Antimicrobial compounds in food packaging. International Journal of Molecular Sciences. 2023;**24**(3):2457

[110] Kuswandi B. Environmental friendly food nano-packaging. Environmental Chemistry Letters. 2017;**15**(2):205-221

[111] Naseer B, Srivastava G, Qadri OS, Faridi SA, Islam RU, Younis K. Importance and health hazards of nanoparticles used in the food industry. Nanotechnology Reviews. 2018;**7**(6):623-641

[112] Jagtiani E. Advancements in nanotechnology for food science and industry. Food Frontiers. 2022;**3**(1):56-82

[113] Zuo J, Feng J, Gameiro MG, Tian Y, Liang J, Wang Y, et al. RFID-based sensing in smart packaging for food applications: A review. Future Foods. 2022;**6**:100198

Chapter 3

Model-Based Experimental Design to Optimize Moroccan Sweet Pepper Cultivation in Nursery: A Comprehensive Study of *Phytophthora capsici* Management Using Experimental Factorial Design

Wafaa Mokhtari, Malika Ablagh, Mimoun Mokhtari, Noureddine Chtaina and Mohamed Achouri

Abstract

This research addresses the challenge of managing *Phytophthora capsici* in nurseries, emphasizing an integrated approach for optimal pepper health employing model-based experimental design. The study employs a factorial design and randomized block experiment to manipulate factors such as pathogen inoculum, Trichoderma treatment, and indoor environmental conditions. The factorial design provides insights into the intricate interactions that influence *Phytophthora capsici* dynamics. By identifying pathogen inoculum thresholds varying the amount [0; 500,000; 1000,000 (CFU/ml)], optimizing *Trichoderma* concentrations giving the range of [0; 1000; 10,000,000 (CFU/ml)], and assessing the impact of environmental conditions, we can enhance our understanding of biocontrol efficacy. The results offer valuable insights for the development of effective and tailored disease management strategies. The study's implications extend to efficient resource utilization identifying optimal environmental conditions; T° > 20°C, RH% < 80%, and biocontrol treatment optimal concentration of 10,000,000 CFU/ml for the development of targeted disease management strategies. This research highlights the importance of experimental factorial design in understanding the complexity of Phytophthora communities applying a more flexible structured approach; blocking factors in 3x2 and 2x2x2x2, to improve the accuracy of treatment effect estimation, the results have significant implications for future research in Phytophthora root rot diseases control and monitoring.

Keywords: phytophthora dynamics, disease management in the nursery, model-based factorial experiment design, empirical, biocontrol

1. Introduction

1.1 Pepper production, and disease management impact

Moroccan sweet peppers, along with other solanaceous vegetable crops, hold a pivotal position as primary target products in horticulture due to their significant dominance in volume to meet global market demands [1]. However, high-yield production potential is challenged by unsustainable management practices and diseases including root rot disease caused by *Phytophthora capsici*, which threatens overall productivity. *Phytophthora capsici* is a significant threat to a variety of horticultural plants, including the Moroccan sweet pepper cultivation. The impact of *Phytophthora capsici* on pepper is influenced by various factors such as inoculum prevalence during and after infection, and environmental conditions. This monocyclic disease is initiated by triggering oospore germination under optimal conditions, such as heavy rains and high soil moisture. The zoospore cycle is critical for infection at different stages of pepper growth, causing seed rotting and damping-off during pre-emergence and contributing to symptoms such as root rot and plant death at later stages [2, 3].

Successful adaptation of root rot disease control in off-season-pepper-nurseries cultivation is particularly focused on environmental conditions control that is monitoring high humidity, and temperature and optimizing soil drainage. Implementing transplanting technology management and monitoring environmental systems in nurseries improves the health of pepper seedlings. Emphasizing the importance of such practices is crucial, not least in addressing the demands of the competitive pepper market. In the integrated management of root rot disease, implementing preventive measures can be more effective creating an inhospitable environment for the inoculum source of *Phytophthora*. One valuable strategy to enhance disease management is the incorporation of beneficial biological control agents (BCAs) in seedling and during transplantation. In nursery settings, it is equally important to take a preventive approach by monitoring and regulating environmental factors [4, 5].

1.2 Importance of model-based experimental design in *Phytophthora* management in nurseries

To prevent disease development, the identification of potential stressors in the nursery is paramount. To systematically study the effects of multiple factors, a well-structured design of experimental units is essential. Within this experimental design, specific combinations are assigned to screen for *Phytophthora capsici* and to select *Trichoderma* isolates with potential biocontrol effect [6, 7].

In the context of biocontrol agents, it is important to consider real-world applicability and empirical approaches and field experiments are crucial in this regard [4, 8]. Still, the integration of model-based with pre-defined criteria into the experimental approach, including the combination of multiple factors as well as their interactions, provides a powerful strategy for understanding and selecting effective biocontrol agents. This integration of correlated structures for *Trichoderma* treatment and other environmental factors enables accurate estimation of plot-to-plot variability [8–10].

Phytophthora capsici poses formidable challenges due to its complex biology and diverse life cycle. The importance of experimental factorial design in disease control strategies is crucial. Here's an exploration of why this approach is critical [6, 9–13].

i. Experimental factorial design allows for the systematic manipulation of multiple factors simultaneously, providing a nuanced view of biocontrol dynamics.

ii. Factorial experiments facilitate the quantification of pathogen inoculum thresholds, which is critical for identifying critical thresholds associated with reduced Phytophthora incidence.

iii. It enables researchers to optimize biocontrol measures by assessing the effectiveness of different *Trichoderma* concentrations.

iv. This allows researchers to study how environmental conditions like temperature and humidity affect the efficacy of biocontrol agents to support the integration of new monitoring tools into biocontrol trials and contribute to the evaluation of their efficacy.

v. It allows researchers to test multiple variables simultaneously, optimizing the use of resources.

vi. It also contributes to the development of customized disease management strategies that are adapted to specific conditions.

In summary, factorial design is a key tool for reducing the incidence of *Phytophthora capsici*. Its ability to dissect complex interactions, account for variability and optimize biocontrol measures positions it as a cornerstone in the development of effective and tailored strategies for managing this challenging pathogen. Indeed, the application of factorial design in horticulture recognizes practical constraints and optimizes resource use in root rot disease management.

To investigate the effectiveness of different *Phytophthora* management strategies, a randomized block design was implemented in a controlled nursery environment. Also, we investigate the complex dynamics of *Phytophthora* inoculum in nurseries using a factorial design. The study aims to elucidate interactions between multiple factors, including pathogen inoculum levels, *Trichoderma* concentrations, and environmental conditions, to optimize root rot disease management [8–10, 14].

2. Material and method

2.1 Obtaining culture of *Phytophthora*

A culture of *Phytophthora capsici* was obtained by isolating Phytophthora mycelium from pepper roots exhibiting blight disease at the IAV CHA greenhouse in Ait-melloul, Morocco. To achieve this, pepper roots displaying clear symptoms of rot were identified and collected. Surface baiting was performed on the selected roots using two different agar media: corn meal agar (CMA) and a modified CMA selective medium (CMAm), following the method proposed by Tsao and Ocana in 1969. The roots were disinfected by immersing root fragments, ranging from 5 to 10 mm, in a 10% solution of sodium hypochlorite. After disinfection, the fragments were rinsed for 1 minute in distilled deionized water to remove any residual disinfectant. Finally, the disinfected root fragments were dried on sterile filter paper to remove excess moisture. The dried root fragments were placed on the surface of corn meal agar medium (CMA) and incubated in the dark at a constant temperature of 21°C for 5 to 7 days [15].

2.2 Preparation of *Phytophthora capsici* zoospores inoculum

To prepare the inoculum of *Phytophthora capsici*, consisting of sporangia and zoospores, the method outlined by Chen and Zentmyer [16] was meticulously followed. The mycelium culture was grown in a V8 agar medium and then carefully cut into 5 to 7 cm pieces under aseptic conditions to serve as the source for inoculum preparation.

To ensure cleanliness, all cut pieces underwent two aseptic washes using sterilized deionized water. The washed mycelium pieces were then placed in sterile glass petri dishes and flooded with 10 ml of sterilized deionized water. The washed mycelium pieces were then placed in sterile glass petri dishes and flooded with 10 ml of sterilized deionized water. Non-sterile soil was added to the water in the petri dishes. The petri dishes were prepared and then incubated at 22°C for 3 days [16].

2.3 Assessment of zoospores release

To evaluate sporangia production after 3 days of incubation, sporangia were observed microscopically. Zoospore release was induced by subjecting the glass petri dishes containing mycelium pieces to a thermal shock. The dishes were placed at 4°C for 30 to 60 minutes, followed by a return to the incubator at 22°C for 15 to 20 minutes. The method of thermal shock, described by Chen and Zentmyer (1970), Dhingra and Sinclair (1985), Guo and KO (1993), and Nielsen et al. (2006), facilitated the release of zoospores [16–19].

2.4 Obtaining Trichoderma subculture

The standard procedure for soil sampling was followed to collect soil samples near Argania roots in the Argania forest, Agadir province, Morocco [20]. The dilutions (ranging from 10^{-2} to 10^{-5}) were spread on a potato dextrose agar (PDA) medium supplemented with ascorbic acid and incubated at room temperature for 1 to 2 weeks. The collected samples were then suspended and diluted. Colonies resembling Trichoderma spp. were isolated from the plated dilutions and transferred to fresh PDA plates. For subculture preparation, 10 µl of sterile saline spore solution was added to each colony. Each spore suspension was then inoculated onto the surface of a fresh 3% malt extract agar (MEA) plate. The subculture rounds were incubated for 2 to 4 days at 28°C. The purified subcultures were grown and maintained on MEA or PDA medium. They were identified at the genus level based on their morphological and cultural characteristics, as described in previous work [21].

2.5 Pepper root-dipping inoculation

For each cultivar, the roots were dipped in a spore suspension of *Trichoderma* species. This study aims to examine the preparation of *Phytophthora* inoculum, inoculation procedures, and disease assessment, with and without *Trichoderma* spp. treatments. Both *Trichoderma* and *Phytophthora* were in suspension and diluted to a fixed concentration in sterile deionized water before inoculation. The root-dipping method was selected as the inoculation technique for this research to facilitate the rapid and direct percolation of *Trichoderma* spore suspensions into the root tissue of host plants. During transplanting, pepper seedling roots were immersed in fungal suspensions. Initially, the roots were dipped in a suspension containing 10^8 conidia/ml of various *Trichoderma* spp. for 15 to 20 minutes. The roots were immersed in a suspension containing 5.10^5 conidia/ml of *Phytophthora* zoospores for 3 to 5 minutes. This process was repeated as described in Ref. [22].

2.6 Disease incidence evaluation

Disease incidence was evaluated by quantifying the percentage of plant units displaying typical symptoms associated with each corresponding pathogen. The disease incidence percentage (DI%) was calculated by dividing the number of plant units with typical symptoms by the total number of plant units and multiplying by 100 (see Eq. (1)).

$$DI\% = \frac{Number\ of\ infected\ plant\ units}{number\ of\ plants} * 100 \tag{1}$$

DI% was evaluated in both above-ground and root units. Symptoms such as yellowing and necrosis in leaves, wilt and/or stunt in shoots, and rots and discoloration in stems were identified and measured. Furthermore, symptoms and signs retrieved from roots, crowns, stems, and leaves were documented for each pathogen. The study examined signs of pathogen components, such as mycelium and propagules, under microscopic observation [23, 24].

2.7 Randomized complete block

2.7.1 Experimental design

Pepper cultivars were grown for 4 weeks in 77 peat trays. Upon transplantation into 3 L pots, seedlings were successively inoculated with the pathogen and the biocontrol. The pots were filled with sterile substrate at a 3:1 w/w peat-to-sand ratio and fertilized using a composition of NPK and oligo-elements (20-20-20 hydrosoluble NPK plus magnesium, iron, and manganese) at a concentration of 250 mg/l.

The experimental design consisted of four randomized complete blocks, each with four replicates in every experimental unit [25]. Each experimental unit was composed of four pots (**Figure 1**).

2.8 Full factorial experimental design

The study employed a full factorial experimental design to investigate the relationships between Phytophthora incidence, biocontrol measures, and environmental monitoring in nursery settings. The methodology comprises two experiments, each focusing on specific factors that influence disease dynamics [8, 10, 14].

2.8.1 Experiment 1

In this first model-based factorial experiment, controlled variables were deliberately selected for both the pathogen and biocontrol agent. The aim was to investigate

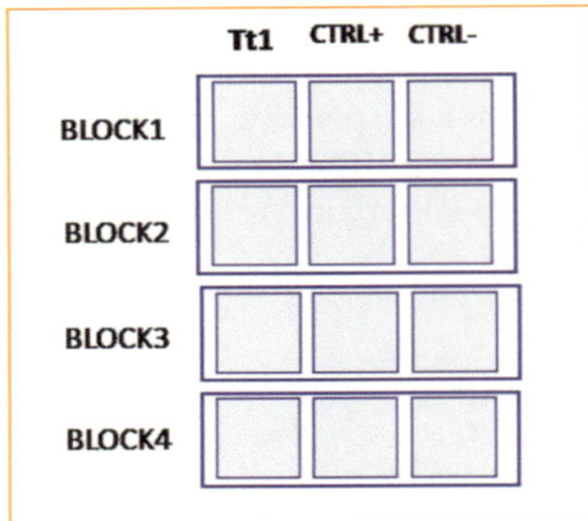

Figure 1.
The experimental design consisted of four randomized complete blocks representing the first initial nine runs, for a total of 16 runs the same design is replicated four times.

the direct correlation between overall DI% and the concentration of the pathogen inoculum, as well as the application of *Trichoderma* treatment. Each factor has three established levels.

The levels of pathogen inoculum concentration were set at 0, 500,000 and 1000,000 colony-forming units per milliliter (CFU/ml), and the levels of Trichoderma treatment were set at 0, 1000 and 10,000,000 (CFU/ml). This design enables a confident evaluation of the impact of different inoculum concentrations and *Trichoderma* treatments on the incidence of *Phytophthora*.

2.8.2 Experiment 2

The second experiment examines the complex interactions among four critical input factors: pathogen inoculum (Pi), biocontrol treatment (Tt), temperature (T°), and relative humidity (RH%) monitoring. A full factorial design is used to assess two levels for each factor, resulting in a comprehensive matrix of experimental conditions (see **Figure 2**). The pathogen inoculum concentration levels are 0 and 1000,000

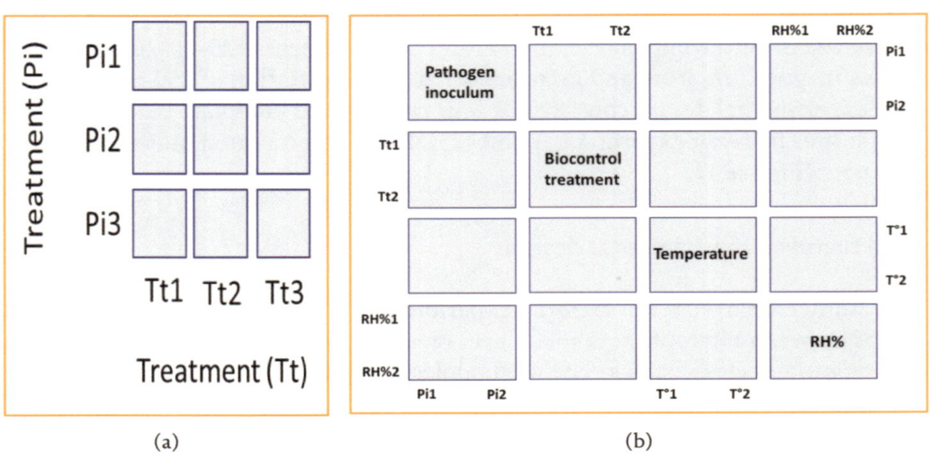

Figure 2.
Blocking factorial in a 3x3 arrangement (a), with four repetitions across four blocks. And (b) blocking factorial in 2x2x2x2 configuration for the combination of all four factors (pathogen inoculum (Pi), biocontrol treatment (Tt), temperature (T°), and relative humidity (RH%) with only one replicate.

(CFU/ml), while the *Trichoderma* treatment levels are 0 and 10,000,000 (CFU/ml). The temperature levels tested were 20 and 40°C, while the relative humidity levels were 30 and 80%.

To ensure robustness, the first design was evaluated in a 3x3 arrangement, with four repetitions across four blocks. The second design employed a more intricate 2x2x2x2 configuration, featuring a single repetition. This design facilitated a thorough examination of both interaction and main effects (refer to **Figure 2(a)** and **(b)** in the result).

3. Results

3.1 *Phytophthora capsici* culture

Phytophthora capsici was confidently isolated and identified using a selective CMA medium. The white mycelium exhibited an alternate diffusible structure on CMA selective media, and microscopic observation revealed non-septate mycelium and characteristic lemon/olive-shaped sporangia structures. Sub-culturing the mycelium on PDA improved the visibility of the alternate structure, making it easier to obtain a pure culture of the pathogen. A culture-based approach successfully identified the fungi at the genus level, highlighting the significance of culture characteristics in identifying fungi based on mycelium and propagule morphology.

3.2 *Phytophthora capsici* zoospore

The artificial infection method was successful in identifying the pathogenicity of *Phytophthora*. Zoospores were effectively produced, as shown in **Figure 3**.

Trichoderma subcultures were successfully developed on PDA, with colonies identified as *Trichoderma* spp. based on their distinct morphological characteristics, such as colony color. The colonies displayed a diverse range of colors, from yellowish to dark green. Pure cultures of *Trichoderma* isolates were successfully obtained after multiple rounds of sub-culturing.

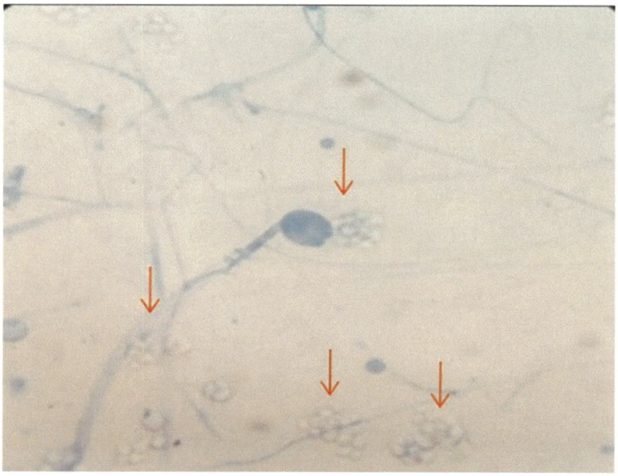

Figure 3.
Zoospores release from sporangia, magnification 400X.

3.3 *Phytophthora* treatment in randomized complete block

The randomized complete block experiment showed that *Trichoderma* spp. root-dipping treatments significantly reduced the incidence of rots in pepper (P = 0.000). The results are summarized in **Table 1**, which shows the treatments and their corresponding DI% values.

Disease incidence in pepper plants was recorded by observing brown to black rots in roots, crowns, or stems. The *Trichoderma* treatment resulted in a DI% of 6.3%, compared to DI% = 100% in CTRL+ and DI% = 0.0% in healthy plant CTRL- (**Figure 4**).

Trichoderma exhibits effectiveness in treating infected pepper plants in each block group within every treatment. However, the random assignment of treatments (Trichoderma) or healthy plants' controls (CTRL-) within each block constrains the exploration of all conceivable combinations of factor levels to a limited set, such as in the control (CTRL-) where there is no inoculation at all and CTRL+ (a high level of

Treatments	DI%*
CTRL+	100% ± 0.0a
CTRL-	0.0% ± 0.2d
Trichoderma	6.3% ± 0.1a

CTRL+: Pepper plant infected with Phytophthora, and CTRL-: non-inoculated healthy plant. The means of four replicates (16 replicates) in four blocks are represented by the standard deviation ±. Significant differences were determined using Tukey's test grouping at p = 0.05 and letters were assigned accordingly.

Table 1.
DI% for pepper plant inoculated with Trichoderma.

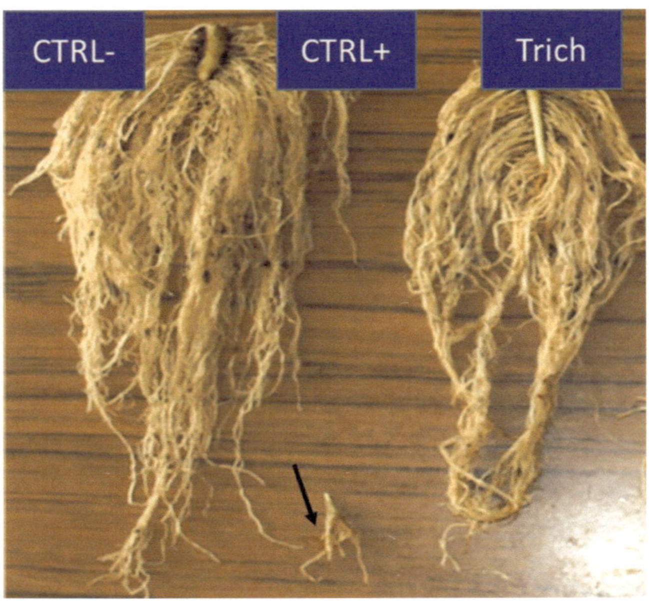

Figure 4.
CTRL-; healthy pepper with no pathogen or Trichoderma *inoculation, CTRL+; pepper root infected with* Phytophthora capsici *inoculum; Trich;* Trichoderma *as biocontrol treatment against* Phytophthora *infection.*

pathogen amount). This design may fall short in capturing interaction effects between factors, as it does not ensure a balanced representation of all possible combinations.

3.4 Deducing Phytophthora treatment in a full factorial design

3.4.1 2-factorial regression analysis

In this factorial experiment, a comprehensive regression analysis was conducted to investigate the effects of Phytophthora inoculum and Trichoderma treatment on disease incidence (DI%). The experimental design included three levels of pathogen inoculum (Pi); 0, 500,000, and 1000,000 CFU/ml and three levels of Trichoderma treatment (Tt); 0, 1000, and 10,000,000. The regression model effectively captures and explains the complex relationship between the studied factors and disease incidence. The coefficient of determination (R-sq) was significantly high at 98.84%, indicating that the model effectively explained a substantial portion of the variability in DI%. Further supporting the model's effectiveness, the adjusted R-squared (R-sq(adj)) stood at 98.50%.

The statistical significance of the model was confirmed through analysis of variance (ANOVA) ($p < 0.001$). All main effects and interaction terms were significant, indicating a substantial influence of both inoculum and treatment factors, as well as their interaction, on DI%.

Eq. (2)'s coefficient reveals the impact of the pathogen inoculum on DI%, showing a significant positive effect at higher inoculum levels (500,000 and 1000,000). Treatment with Trichoderma at levels 0 and 1000 had a positive influence on DI%, while treatment at level 10,000,000 had a notable negative impact on DI%.

$$\begin{aligned}39{,}369 - 39{,}37\text{Pinoculum\_0} + 18{,}24\text{Pinoculum\_500000} + 21{,}13\ \text{Pinoculum\_1000000} +\\ 18{,}30\ \text{Ttreatment\_0} + 17{,}13\ \text{Ttreatment\_1000} - 35{,}43\ \text{Ttreatment\_10000000}\end{aligned} \tag{2}$$

(refer to the supplementary document for the complete regression equation).

3.4.2 Main and interaction effect

The results in the main effect of factors combination provide a comprehensive understanding of the relationships between *Phytophthora capsici* inoculum as a factor influencing DI% and the effective performance of *Trichoderma* treatment as a biocontrol (**Figure 5**). The interaction terms further clarified the combined effects of pathogen inoculum and treatment on DI% (**Figure 6**).

3.4.3 Multifactorial (4-factor) design 2x2x2x2

The study demonstrated that the multifactorial design, characterized by a 2x2x2x2 configuration, effectively explains the variation in DI% with a remarkable R-squared value of 100.00%. The adjusted R-squared, which considers the number of predictors, remained high, affirming the model's robustness.

The analysis of variance (ANOVA) for the general factorial regression model yielded significant results, highlighting the collective impact of factors such as Phytophthora inoculum (Pi), Trichoderma treatment (Tt), temperature (T°), and relative humidity (RH%) on the response variable (DI%). The observed variation

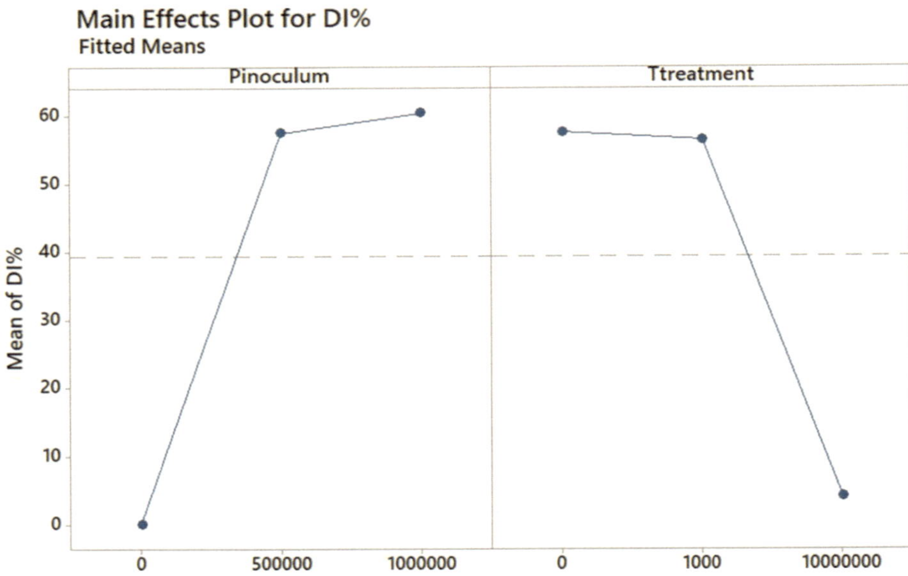

Figure 5.
Main effect plot of the Phytophthora inoculum *(Pinoculum) and* Trichoderma *biocontrol (Ttreatment) on DI% represented by a non-horizontal line with a higher DI% mean in 500,000 and 1000,000 CFU/ml for (Pinoculum) and 1000 and 0 CFU/ml for (Ttreatment). The slope inclination of each line shows the magnitude of the DI% response.*

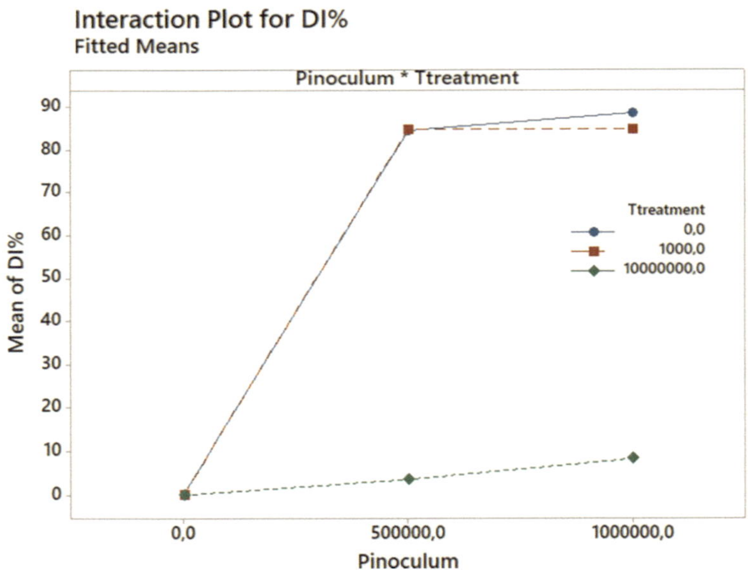

Figure 6.
The graph of interaction effect shows the combined effects of Phytophthora inoculum *(Pinoculum) and* Trichoderma *treatment (Ttreatment) on DI% which is in harmony with the main effect plots result. Notably, different intercepts are observed at all levels, indicating different effects of Pinoculum and Ttreatment on DI%. The highest levels 500,000 and 1000,000 CFU/ml of* Phytophthora inoculum *(Pinoculum) are both associated with the highest mean of DI%. Whereas, the lowest levels of* Trichoderma *treatment (Ttreatment) 1000 and 0 CFU/ml are the ones associated with the highest mean of DI%.*

indicates that the linear main effects of Pi, Tt, T°, and RH% are all significant ($p < 0.05$), signifying the individual influence of each factor on DI%.

The main effects plot shows that the magnitude of the highest and lowest mean for each factor is strongly high, as observed by the slope lines. The impact on the response variable (DI%) is more pronounced with steeper slopes. All pairs of factors (Pi*Tt, Pi*T°, Pi*RH%, Tt*T°, Tt*RH%, T°*RH%) show significant interactions, highlighting the combined effects when considering two factors simultaneously. The interaction plot illustrates how the joint influence of two factors varies (**Figure 7**).

Interpreting the slopes and patterns in **Figure 7** provides valuable insights into the combined effects of factors influencing DI%. For optimal control of DI%, it is recommended to monitor the temperature at its highest (40°C) and maintain the relative humidity at its lowest (30%). Even at the most favorable temperature condition of 20°C for *Phytophthora*, maintaining a relative humidity of 30% minimizes DI%.

The study found significant three-way interactions (PiTtT°, PiTtRH%, PiT°RH%, TtT°RH%) and a significant four-way interaction (PiTtT°*RH%), highlighting the collective impact of all four factors when considered together. It is important to note that the influence is primarily attributed to specific combinations, as shown in the Pareto chart in **Figure 8**. The Pareto chart identifies the 'vital few' as portions A and B, which significantly repress DI%. The cumulative line, crossing 80% with A, B, C, and D, underscores the substantial contribution of these major combinations to the overall impact on DI% evolution.

3.4.4 Multifactorial regression equation analysis

To ensure more effective solutions through the representation of the relationship between different factors and the predicted (DI%) response, it is crucial to prioritize

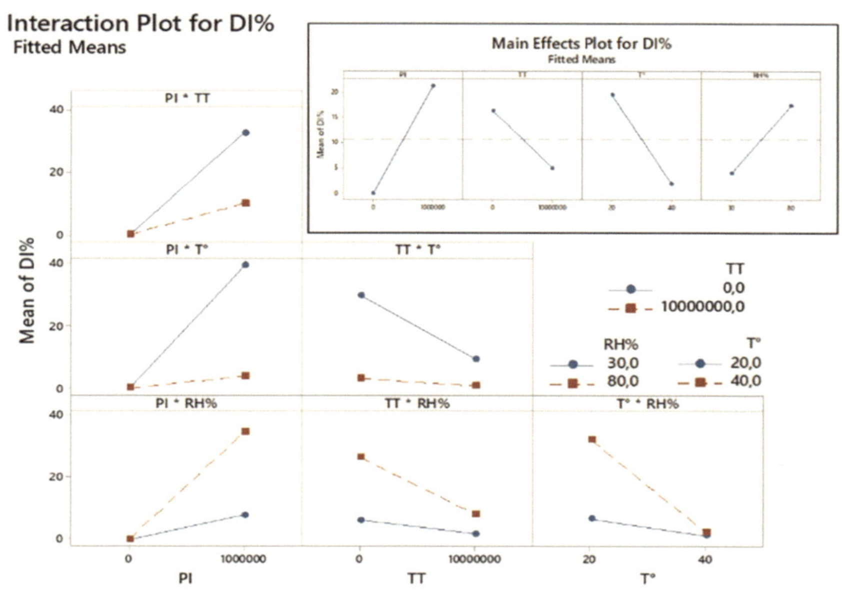

Figure 7.
*Interaction effect plot of every pair of factors (Pi*Tt, Pi*T°, Pi*RH%, Tt*T°, Tt*RH%, T°*RH%) with simultaneous influence on the variable response (DI%) with the main effect plot highlighting the magnitude of response (DI%) at different factor level.*

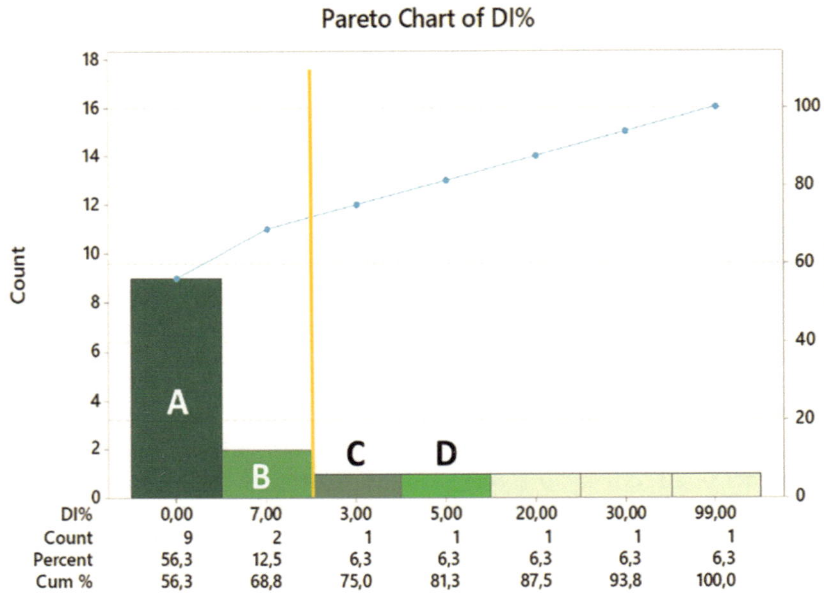

Figure 8.
*Patero chart of portion with factors' combinations with significant impact on DI%, adjacent **Table** 2 represents the factors' combinations with a succession (Pi*Tt*T°*RH%) representing the portion A, B, C, and D.*

Portion	Factors' combinations
A	0*10000000*20*30
	0*0*40*80
	1000000*10000000*40*30
	0*10000000*40*30
	0*10000000*40*80
	0*0*20*80
	0*10000000*20*80
	0*0*20*30
	0*0*40*30
B	1000000*0*40*80
	1000000*10000000*20*30
C	1000000*10000000*40*80
D	1000000*0*40*30

Table 2.
*Factors' combinations with a succession (Pi*Tt*T°*RH%) representing the portions A, B, C, and D.*

coefficients with the highest impact. The examination of the interaction reveals coefficients of −3.563 (Pi*Tt*T°*RH%; 1000,000*10,000,000*40*30) and 5.687 (Pi*Tt; 0*10,000,000), indicating that the influence of *Phytophthora* inoculum (Pi) and *Trichoderma* treatment (Tt) on DI% varies depending on the levels of other factors, such as temperature (T°) and relative humidity (RH%). However, caution is

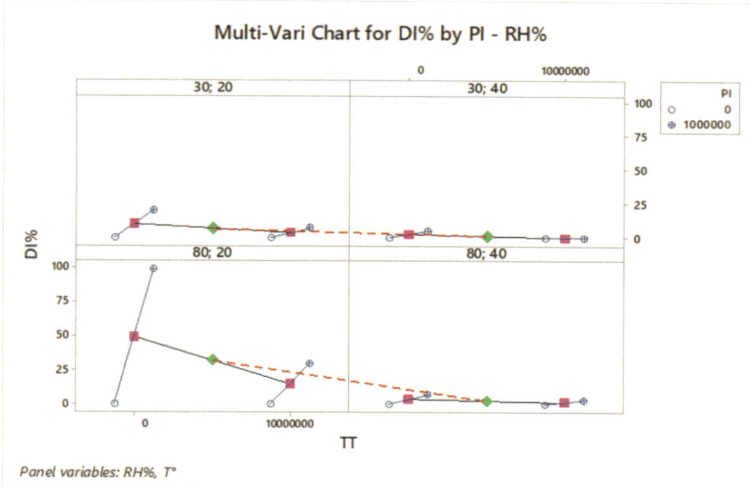

Figure 9.
Multivariate chart: Factors' combinations and dynamic of significant impact on DI%.

necessary when interpreting significant interactions. For instance, when examining the Pi*Tt interaction with coefficients of −5.687 (Pi*Tt; 0*0) and 3.563 (Pi*Tt*T°*RH %; 1000,000*0*40*30), it is essential to acknowledge that combinations of factors and their interactions can result in subtle effects.

Coefficients may not apply universally to every situation, and general trends should be interpreted with caution. As shown in **Figure 9**, the biocontrol effect at this level may interact with other conditions.

4. Discussion-conclusions

This research highlights the importance of using experimental factorial designs to study Phytophthora communities in nurseries, especially when resources are limited and greenhouse environments are controlled. The study demonstrates the crucial role of experimental factorial design in effectively managing the complexities inherent in the Phytophthora population. To comprehensively understand the interactions and variations that influence disease dynamics, researchers manipulate multiple factors simultaneously. In real-world conditions, external variations can impact study outcomes. Therefore, the incorporation of multiple-blocking factors, such as pairs of columns and rows, proves instrumental in controlling variability. This enhancement in precision enables a more accurate estimation of treatment effects, which is crucial for drawing confident conclusions about the true impacts of our treatments [6, 8, 9].

The experimental units are grouped into rows in a randomized complete block design, as shown in **Figure 1**, based on their similar characteristics or conditions within the greenhouse. The treatments, including Trichoderma, CTRL+, and CTRL-, are randomized within each block and represented in columns. This means that any observed differences in DI% are more likely to be due to the treatments (Trichoderma, CTRL+, and CTRL-) rather than block-specific factors [8, 25].

Blocking in factorial design enhances the precision of treatment effect estimation by minimizing the impact of external variations, similar to a completely randomized block design. However, factorial designs explicitly address interactions between factors, providing a more nuanced understanding of how different elements combine to affect DI%. This is especially important when multiple factors are involved, making it more efficient in terms of resource utilization compared to a completely randomized block design.

In a two-factor factorial design, such as pathogen inoculation and *Trichoderma* treatment, the main effects of each factor and their interaction are simultaneously investigated to understand how they may work together to influence DI%. The correlation between Phytophthora inoculum levels (500,000 and 1000,000 CFU/ml) and DI% is strong. This has important implications for biocontrol strategies, as it suggests that the effectiveness of biocontrol measures may be limited at these elevated inoculum levels. Therefore, treatment concentrations may need to be reassessed at higher concentrations. It is worth noting that the potential failure of biocontrol in suppressing root rot disease becomes more prominent when targeting inoculum concentrations below 1000 CFU/ml.

In 2x2x2x2 factorial design, the systematic variation of conditions is extended to include additional factors such as temperature and relative humidity. Monitoring relative humidity at its lowest point (30%) is effective in minimizing DI%, particularly in the context of favorable temperature conditions (<20°C) for Phytophthora. This provides insights into how these climate variables may influence disease dynamics, along with pathogen and treatment effects. Moreover, similar to the two-factor scenario, blocks in this factorial configuration can be defined based on sources of variation that are not the primary focus of the study but might impact DI%. For example, different sections of the greenhouse could be considered, which is a crucial aspect for developing targeted management strategies [8, 9, 14].

4.1 Implications for disease management in nurseries

In summary, the systematic planning of experiments with model-based factorial design optimizes resource utilization in nurseries. This is particularly important as characterizing *Phytophthora* communities can be labor-intensive. By facilitating controlled variation of multiple factors influencing inoculation or infection methods, experimental conditions that closely mimic real-world scenarios within nurseries are created through factorial design, providing more realistic insights into *Phytophthora* dynamics. That is, factorial design enhances the reproducibility of experiments, which is a crucial aspect in nurseries where consistency is paramount [6, 8, 10].

Additionally, a structured approach guides the implementation of cost-effective strategies, particularly when leveraging newer technologies. On the other hand, effective nursery management requires consideration of various factors beyond the conventional focus on temperature and humidity. While these factors are undoubtedly important, a holistic approach to greenhouse performance demands attention to dynamic environmental changes, energy efficiency, comprehensive control systems, water management strategies, and disease prevention caused by Oomycetes and in our case *Phytophthora*. Greenhouse efficiency depends on an integrated approach that goes beyond simplistic factorial models. It is important to recognize the multifaceted nature of factors that influence protected vegetable crop health and its overall production [14].

In conclusion, this study emphasizes the crucial role of model-based experimental factorial design in understanding the complexities of Phytophthora communities.

The insights gained from this structured approach make a significant contribution to disease management strategies in nurseries, demonstrating the potential for impactful and resource-efficient research in controlled environments. The implications of our findings are significant for future research in *Phytophthora* control and monitoring, as we have developed the study to establish and model the stage for such endeavors using factorial design.

Author details

Wafaa Mokhtari[1]*, Malika Ablagh[2], Mimoun Mokhtari[2], Noureddine Chtaina[3] and Mohamed Achouri[1]

1 Plant Protection Department, Agronomic and Veterinary Institute Hassan II, Campus Agadir, Agadir, Morocco

2 Horticultural Department, Agronomic and Veterinary Institute Hassan II, Campus Agadir, Agadir, Morocco

3 Plant production, Protection and Biotechnology Department, Agronomic and Veterinary Institute Hassan II, Campus Rabat, Rabat, Morocco

*Address all correspondence to: w_mokhtari@yahoo.fr

References

[1] Bickelmann U. UNECE standard for sweet peppers FFV-28 (2009). UNECE explanatory brochure for sweet peppers Agenda item 4(f) and Agenda item (7). UNECE and OCED Meeting Report. 2009. Available from: https://unece.org/DAM/trade/agr/promotion/2010_CapeTown/TechnicalSessions/Sweet%20peppers/SweetPeppers_PresentationOfThe Standard.pdf

[2] Nsiah, Frimpong B, Oteng Ampadu S, Oppong A, Nunoo I, Brobbey L. Phytophthora Diseases Prevalence, its Effects and Controls in Ghana. In: Mohamed H, Abdulkhair W, editors. Agro-Economic Risks of Phytophthora and an Effective Biocontrol Approach. London, UK: IntechOpen; 2021. DOI: 10.5772/intechopen.99130

[3] Zaki N, Hakmaoui A, Ouatmane A, Fernandez-Trujillo JP. Quality characteristics of Moroccan sweet paprika (Capsicum annuum L.) at different sampling times. Food Science and Technology International. 2013;**33**(3):577-585

[4] Lavender DP. Plant physiology and nursery environment: Interactions affecting seedling growth. In: Duryea ML, Landis TD, Perry CR, editors. Forestry Nursery Manual: Production of Bareroot Seedlings. Forestry Sciences. Vol. 11. Dordrecht: Springer; 1984. DOI: 10.1007/978-94-009-6110-4_14

[5] Hashemi M, Tabet D, Sandroni M, Benavent-celma C, Seematti J, Andersen CB, et al. The hunt for sustainable biocontrol of oomycete plant pathogens, a case study of Phytophthora infestans. Fungal Biology Reviews. 2022;**4**:53-69

[6] Moreira T, Groot Koerkamp P, Janssen A, Stomph T-J, van der Werf W. Breaking the mould: Developing innovative crop protection strategies with reflexive. Interactive Design Agricultural Systems. 2023;**210**:103727

[7] Nur ZA, Nour BA. Biological functions of Trichoderma spp. for agriculture applications. Annals of Agricultural Sciences. 2020;**65**(2):168-178

[8] Barlow ND. Models in biological control: A field guide. In: Hawkins BA, Cornell HV, editors. Theoretical Approaches to Biological Control. Vol. 1999. Cambridge: Cambridge University Press; 1999. pp. 43-68. DOI: 10.1017/CBO9780511542077.005

[9] Godfray HCJ, Waage J. Predictive modeling in biological control: The mango mealy bug (Rastrococcus invadens) and its parasitoids. Journal of Applied Ecology. 1991;**28**:434-453

[10] Singh R, Chai F-S, Das A. Optimal two-level choice designs for any number of choice sets. Biometrika. 2015;**102**(4):967-973. DOI: 10.1093/biomet/asv040

[11] Evans JA, Davis AS, Raghu S, Ragavendran A, Landis DA, Schemske DW. The importance of space, time, and stochasticity to the demography and management of Alliaria petiolata. Ecological Applications. 2012;**22**:1497-1511

[12] McEvoy PB. Theoretical contributions to biological control success. BioControl. 2018;**63**:87-103. DOI: 10.1007/s10526-017-9852-6

[13] Mohd Ridza PNY, Ya'acob A, Zainol N, Mortan SH. Application of two-level factorial design to study the microbe growth inhibition by pineapple leaves juice. IOP Conference Series:

Materials Science and Engineering. 2020; **736**:022011. DOI: 10.1088/1757-899X/736/2/022011

[14] Castilla N. Regulation and Control System: Computer Climate Management in Greenhouse Technology and Management. London, UK: British Library, Centre for Agriculture and Biosciences CABI International; 2013. pp. 204-217

[15] Tsao HP, Ocana G. Elective isolation of species of Phytophthora from natural soil on an improved antibiotic medium. Nature. 1969;**223**:636-638

[16] Chen W, Zentmyer GA. Production of sporangia by Phytophthora cinnamomi in axenic culture. Mycologia. 1970;**62**:397-402

[17] Dhingra OD, Sinclair JB. Basic Plant Pathology Methods. Boca Raton, Florida, USA: CRC Press, Inc; 1985

[18] Guo LY, Ko WH. Two widely accessible media for growth and reproduction of Phytophthora spp and Pythium spp. Mycologia. 1993;**59**: 2323-2325

[19] Nielsen CJ, Ferrin DM, Stanghellini ME. Efficacy of biosurfactants in the management of Phytophthora capsici on pepper in recirculating hydroponic systems. Canadian Journal Plant Pathology. 2006; **28**:450-460

[20] Simmons K. Soil Sampling. Standard Procedure Operation. Georgia: Environmental Protection Agency Laboratory Services & Applied Science Division Athens; 2023. pp. 1-30

[21] Mokhtari W, Chtaina N, Halmschlager E, Volgmayr H, Stauffer C, Jaklitsch W. Molecular identification Trichoderma with tef1 & evaluation of their antagonistic potential. Revue Marocaine de Sciences Agronomiques et Vétérinaires. 2017; **5**(3):248-254

[22] Widmer TL. Screening Trichoderma species for biological control activity against Phytophthora ramorum in soil. Biological Control. 2014;**79**:43-48

[23] Benson DM, Baker R. Epidemiology of Rhizoctonia solani pre-emergence damping-off of radish: Influence of pentachloro-nitrobenzene. Phytopathology. 1974;**64**:38-40

[24] Campbell CL, Neher DA. Estimating disease severity and incidence. In: Campbell CL, Benson DM, editors. Epidemiology and Management of Root Diseases. New York: Springer Verlag; 1994. pp. 117-147

[25] Clewer AG, Scarisbrick DH. Practical Statistics and Experimental Design for Plant and Crop Science. Chichester, UK: Wiley; 2001

Chapter 4

Disease Diagnosis in Tea (*Camellia sinensis* (L.) Kuntze): Challenges and the Way Forward

Ganga Devi Sinniah and Niranjan Mahadevan

Abstract

Derived from the *Camellia sinensis* (L.) Kuntze plant, tea is the most widely consumed natural beverage in the world. Tea is a perennial woody plant. Monoculturing tea on a large scale makes it susceptible to many perennial and seasonal diseases. The leaves, stems, and roots of tea plants are susceptible to fungal, bacterial, and viral pathogens. Tea is predominantly grown in Asian and African regions; hence, conventional methods including symptomology and signs, and microbiological and microscopic methods are mainly used in disease diagnosis and pathogen identification. Accurate and rapid identification of diseases and pathogens is imperative for the sustainability of tea plantations. Thus, the technological advancement in plant disease diagnosis also embraces the global tea industry. This chapter discusses new technological advances in tea disease diagnosis, focusing on molecular biological methods, whole genome sequencing, and remote sensing and image analysis methods. Further, it highlights the challenges in disease diagnosis as a perennial woody plant and pins down available opportunities that could be successfully adopted to overcome the issues.

Keywords: disease diagnosis, image analysis, PCR, symptoms, whole genome sequencing

1. Introduction

Tea (*Camellia sinensis* (L) O. Kuntze), belonging to the Family Theaceae, is a cash crop extensively cultivated in regions spanning Asia, Africa, and South America. Prominent tea-producing countries include India, China, Kenya, Sri Lanka, Vietnam, and Indonesia [1]. The global tea trade holds a substantial value of approximately USD 9.5 billion, serving as a crucial source of export earnings. Young succulent leaves are the harvestable part of the tea plant being used to manufacture the second most consumed nonalcoholic healthy beverage in the world. Tea production faces a significant challenge from microbial pathogens. These pathogens not only directly lead to crop losses but also shorten the plant's lifespan, impair the quality of processed tea, and necessitate increased investments in disease management. Annual crop losses ranging from 10 to 30% have been documented due to microbial diseases in tea.

Diseases affecting tea plants are categorized as leaf or foliar diseases, stem diseases, and root diseases. A variety of pathogens such as fungi, bacteria, viruses, and algae are responsible for causing diseases in tea plants. However, fungal diseases pose substantial issues in major tea-producing nations, and about 350 fungal diseases are associated with tea plants [2]. **Table 1** summarizes the most common diseases of the tea plant.

Diseases and causative organism/s	Symptoms	Signs
Leaf disease		
Blister blight *Exobasidium vexans*	• Initial symptom is the emergence of pale-yellow translucent spots. • These spots transform into blisters leading to the formation of an indentation. • Subsequently, the blisters undergo necrosis. • Blister on tender stems produce distorted stems and shoot dieback.	• Blisters are characterized by a white-colored hymenium on the swollen surface. • Basidiospores are produced at the tips of basidia, located outside the leaf tissues.
Gray blight *Pestalotiopsis* like species	• Appears as small, brown-colored spots that gradually enlarge and turn gray. • The margins of the lesion retain their brown color, displaying concentric rings.	• Black acervuli can be observed on the lesion.
Brown blight *Colletotrichum* spp.	• Appear as water-soaked, yellowish-green lesions that subsequently enlarge and become deep brown necrotic lesions. • The lesion margins may display a yellowish-green tint with concentric markings.	• Black acervuli can be observed.
Red rust (Algal disease) *Cephaleuros parasiticus*, *C. mycoidea*)	• Algae manifest as red spots (5–7 mm) on leaf, imparting a rusty appearance. • Spots can become lichenized and turn in to white in color. • The leaves take on a variegated appearance (yellow or white). • Infected stems become more rigid and produce longitudinal cracks.	• Rusty appearance is produced by the upright red filamentous structures carrying sporangia laden with spores.
Bacterial leaf and shoot blight *Pseudomonas syringae* pv. *theae*, *Acidovorax avenae*	• Water-soaked spots appear first in young leaves. • As disease progresses, spots merge together into large patches, killing the leaves. • Symptoms in young shoots become visible after about 2 weeks of leaf infection.	• Bacterial streaming under microscope.
Viral diseases Tea plant necrotic ring blotch virus (TPNRBV) and tea plant line pattern virus (TPLPV)	• Leaf discoloration including albino and chlorina. • Necrotic ring blotch can be seen. • Premature defoliation of leaves may occur. • Weaken growth.	
Tea white scab disease *Phyllosticta theaefolia*, *Elsinoe leucospila*	• Appear as small red spots with a grayish white or grayish brown center. • As disease progresses, the spots become Bird's eye shape, round necrotic spots. • Under favorable conditions, spots cover the entire leaf and cause deformity and death.	

Diseases and causative organism/s	Symptoms	Signs
Tea leaf spot *Didymella segeticola* var. *camellia* (Basionym: *Phoma segeticola*)	• Appear as small pinhead-sized spots. • Black-brown color spots can reach up to 1.2 mm in diameter. • Leaf growth can be decelerated.	
Alternaria longipes	• Appear as round (4–6 mm in diameter), brown color spots. • The center of the spot will turn into white in color.	
Epicoccum nigrum	• Spots first appear at the margins. • Gradually expand and become irregularly shaped, dark brown color, necrotic lesion.	
E. layuense	• Small brown and necrotic spots on young and mature leaves. • Spots later become large brown lesions.	
E. sorghinum	• Light or dark brown color, irregular spots appear at the margin. • As diseases progress necrosis at the initial infection site can be seen.	
Lasiodiplodia theobromae	• Initially chlorotic spots appear that gradually become large irregular brown spots. • Spots cover entire width of the leaf with bark brown or black center. • Finally, dried leaves are dropped.	
New leaf blight disease *Muyocopron laterale*	• Reddish brown lesions at the leaf margin.	• Sometimes black ascocarp can be seen on leaf surface.
Arthrinium arundinis	• Appear as gray-brown dots. • Dots expand to produce irregularly shaped dark brown color lesions (3–4 cm).	
Alternaria alternata	• Grayish brown patches first appear around tips and margins of young leaves that later expand toward the midrib. • Leaf curl, death of leaves, and defoliation.	
Stem diseases		
Stem and branch canker *Botryosphaeriaceae* spp. (formally *Macrophoma theicola*)	• At its onset, an oval, sunken, dark lesion appears that causes bark to decay and detach from the wood. • Callusing around the lesion's edges forms thickened bark areas known as cankers which may gradually expand and encircle the stem. • Under severe infection, leaf yellowing, necrosis, dieback, and eventual leaf shedding can be seen.	• Perithecia, varying in color from ash to black, are infrequently sighted as the disease progresses to more advanced stages such as wood decay.
Collar canker and die back *Phomopsis theae*, *Fusarium solani* species complex	• Cracks and bark peeling occur in the collar region, either at the soil level or just below it. • Heavy flowering and fruiting are commonly followed by canopy thinning. • Progressive dieback of branches results in bare bushes.	• Under damp and moist conditions, reddish, tiny superficial perithecia are visible in bushes infected by *F. solani* species complex.

Diseases and causative organism/s	**Symptoms**	**Signs**
Root disease		
Red root disease *Poria hypolateritia* (current name – *Ceriporiopsis hypolateritius*), *Ganoderma philippine*	• Infected roots exhibit a mottled appearance. Over time, both the root surface and inner tissues turns through colors of white, red, and eventually brick red. • With time, the infected roots disintegrate into a soft pulp. • Leaf yellowing, wilting and dieback.	• A fine layer of white mycelium and white mycelial cords are noticeable between the wood and the bark and on the root surface, respectively. • As time progresses, these white mycelial cords turn in to red and fuse together, giving rise to a network that spreads across the root.
Brown root disease *Phellinus noxius*	• The roots are encrusted with a compact mass of soil and small stones that are difficult to remove by washing or rubbing. • As disease progress, the inner tissues of the roots decay, resulting in the formation of a honeycomb-like structure. • Leaf yellowing, wilting and dieback.	• White to brown woolly masses of mycelia are found attached to the encrusted mass of earth. • These mycelial masses gradually turn in to dark brown in color.
Black root disease *Rosellinia arcuate*, *R. bunoides*	• Gray to black coloration is observed on the root surface, extending to the collar region and above in some cases. • The stem's bark near the soil surface may be found dead, spanning a length of 7.5–10 cm. • Leaf yellowing, wilting and dieback.	• White to black strands of mycelia firmly adhere to the roots, forming a loose cobweb-like mass. • Numerous small, white, star-shaped mycelial structures develop on the wood's surface beneath the bark.
Armillaria root rot *Armillaria* spp.	• Premature flowering and somewhat stunted growth are evident. • Longitudinal cracks are typically present above the soil level on the collar, as well as on the taproot and lateral roots. • Leaf yellowing, wilting, and dieback.	• Creamy white, fan-shaped mycelia can be seen under the bark. • Additionally, dark brown to black rhizomorphs with a white interior develop beneath the bark. • Clusters of 5–21 basidiomata can be seen.
Charcoal root disease *Ustulina zonata U. deusta*	• Black, irregular double lines are visible in the wood. • This is often accompanied by leaf yellowing, wilting, and sudden death of the bushes.	• White color, large, fan-shaped patches of mycelium are seen on wood surface upon the removal of bark.

Table 1.
Use of symptoms and signs in the diagnosis of tea diseases.

1.1 Leaf diseases

Blister blight (*Exobasidium vexans*), brown blight or anthracnose (*Colletotrichum* species complex), gray blight (*Pestalotiopsis*-like species), and bacterial blight (*Pseudomonas syringae* pv. *theae* and *Acidovorax avenae*) are the major foliar diseases of tea. The diseases are problematic in all the tea-growing countries in Asia. Gray blight is problematic in African tea-growing countries as well. The crop loss recorded from blister blight and gray blight ranges from 20 to 50% and 10 to 20%, respectively. The extent of crop loss caused by the other diseases is considerable, even though exact figures are unavailable. In addition to crop loss, leaf diseases impact the quality of the final product by altering the biochemical constitution of the leaves. Foliar diseases become problematic generally under cool and humid weather conditions.

1.2 Stem diseases

Cankers are widely reported stem diseases in tea, prevalent in almost all tea-growing countries. The pathogen/s attack twigs and stems, kill or griddle branches, and ultimately kill the entire bush. Cankers on the stems and or branches are referred to as stem and branch cankers. Cankers that originate from the collar region near the soil surface are termed collar cankers (**Figure 1I–J**). *Botryosphaeriaceae* fungi *Lasiodiplodia theobromae*, *Botryosphaeria dothidea*, and *B. mamanae* are reported to cause stem and branch cankers. *Phomopsis theae* (*Diaporthe*) and *Fusarium solani* species complex have been identified as pathogens that cause collar cankers. Cankers are persistent and highly influenced by soil conditions, dry weather, plant nutritional status, and agronomic practices. Hypoxylon wood rot, caused by *Nemania diffusa*, and thorny stem blight, caused by *Tunstallia aculeata* (formally known as *Aglaospora aculeata*) are two other less common stem diseases.

Figure 1.
Symptomology of tea diseases. (A) Blister blight, (B) brown blight, (C) gray blight, (D) red rust disease, (E) Phyllosticta leaf blight, (F) red root disease, (G) black root disease, (H) brown root disease, (I) stem and branch canker, and (J) collar canker.

1.3 Root diseases

Root rots are widespread among tea-growing countries. Root infections result in the sudden death of branches or entire bushes. Some root infections lead to the gradual death of bushes following fruiting, flowering, and defoliation. *Armillaria* spp. *Poria hypolateritia* (*Ceriporiopsis hypolateritia*), *Rosellinia arcuata* (*Dematophora arcuata*), *Phellinus noxius* (*Pyrrhoderma noxium*), and *Ustulina zonata* (*Kretzschmaria zonata*) are major root pathogens in tea. The symptoms of infection expressed on the root by each of the pathogens are unique (**Figure 1F–H**).

Accurate disease diagnosis, identification of the pathogen, and disease quantification are crucial steps at the forefront of effective disease management. Early diagnosis assumes paramount importance as it enables timely intervention and targeted treatment strategies. Furthermore, accurate estimation of disease incidence and severity, along with a comprehensive study of the detrimental impacts of diseases on both harvest quality and quantity, has gained significant importance in various facets of crop cultivation. This chapter summarizes the novel and advanced approaches now in use in disease diagnosis and pathogen identification and quantification in tea and sheds light on the general approaches in this area.

2. Techniques used in tea disease diagnosis

2.1 Symptomology and signs

Symptomology, a fundamental and oldest tool in plant disease diagnosis, involves the observation and interpretation of visible alterations in plant appearance caused by pathogens. By closely examining leaf, stem, and root discoloration, lesions, deformities, wilting, the presence of pathogen structures, and other abnormalities, such as cankers, rots, etc., pathologists have long been able to diagnose diseases efficiently.

The symptomology and signs of various widespread tea diseases are well-documented (**Table 1** and **Figure 1**). Yet, tea disease diagnosis is frequently complicated by overlapping symptoms, leading to confusion. For example, gray, brown, and black lesions on the leaf surface are defined symptoms of gray blight, brown blight, and black blight diseases, respectively. Nevertheless, the subjective nature of color recognition can introduce confusion in disease diagnosis, particularly considering the absence of distinct demarcations among these three colors. Further challenges arise with newly reported leaf spot diseases that often exhibit similar symptoms [3, 4]. Diagnosing stem and root diseases in tea is exceptionally challenging, especially during the early stages of infection. Symptomology and signs of these diseases are often concealed beneath the densely covered tea canopy and the soil, respectively. Even when diseases progress, foliar symptoms tend to resemble one another, including leaf yellowing, necrosis, and branch and bush dieback. Consequently, meticulous observation of symptomology and signs on stems and roots often requires the uprooting or removal of branches, highlighting the difficulty of diagnosis. This process involves labor-intensive field operations. On the other hand, the distinct symptoms that are being used to diagnose root diseases appear at later stages of disease progression. Hence, early diagnosis using this method is extremely challenging.

2.2 Microbiological and microscopic methods

Microbiological and microscopic methods commonly involve the collection of samples from infected plant tissues, followed by culturing them on specialized microbiological media to isolate and identify the target pathogen(s). This process relies on observing the culture morphology and characteristics of asexual and sexual structures under a microscope for accurate identification. Additionally, microscopic methods can be employed to study infected plant tissues, focusing on identifying tissue deformities and the presence of pathogen structures within the tissues.

Culturing and microscopy for correct diagnosis hold significant importance in tea pathology. There are scenarios where a single disease or closely resembling symptoms can be caused by more than one species of pathogens. For instance, multiple species of *Colletotrichum* and *Pestalotiopsis* have been identified as pathogens of brown blight and gray blight diseases. Leaf spot disease can be caused by *Didymella segeticola*, *Alternaria longipes,* and different species of *Epicoccum* [5–7]. Several species within the *Botryosphaeriaceae* family have been identified as pathogens causing stem and branch cankers in Sri Lanka [8]. Collar canker disease can be caused either by *P. theae* or *F. solani* species complex [9]. *Poria hypolateritia* (currently classified as *Ceriporiopsis hypolateritius*) and *Ganoderma philippine* have been documented as the causes of red root disease. Hence, in the absence of well-developed novel technologies, accurate diagnosis of most tea diseases relies heavily on microbiological and microscopic methods.

Since the majority of fungal tea pathogens are necrotrophs, they can be cultivated on commonly used microbiological media like potato dextrose agar (PDA). These pathogens thrive under standard environmental conditions, displaying specific cultural morphologies that are easily recognizable to skilled pathologists. For example, *P. hypolateritia*, *P. noxius*, and *R. arcuata* yield white color colonies on PDA. Over time, these colonies transform into milky white, yellow to brown, and black colors, respectively. Such distinctive color changes provide valuable cues for identification up to the genus level (**Figure 2**).

Light microscopy has been used for disease diagnosis in tea since the blister blight pathogen was first identified in 1898. Fluorescence microscopy and electron microscopy have been used to study the infection process of *E. vexans* [10] and to study the difference between *E. vexans* infected and healthy tea leaves [11]. Fluorescence microscopy using calcofluor white and wheat germ agglutinin stains and scanning electron microscopy of *E. vexans* inoculated tea leaves has been successfully used as a tool to study and diagnose the germination and infection stages of the pathogen [10].

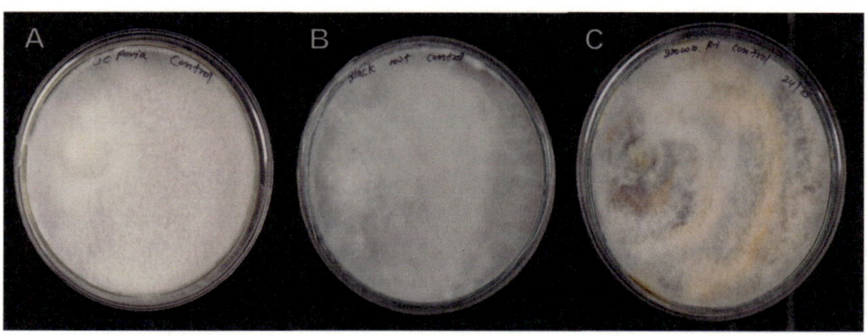

Figure 2.
Culture morphology (upper surface of 10 days old culture) of tea root pathogens (A) Poria hypolateritia (red root disease), (B) Rosellinia arcuata (black root disease), and (C) Phellinus noxius (brown root disease).

2.3 Molecular biological methods

Molecular biological methods in plant disease diagnosis involve the analysis of genetic material, such as DNA or RNA, to identify plant pathogens and eventually diagnose diseases. These techniques have gained popularity in plant pathology due to their accuracy and specificity. Common molecular biological methods used for plant disease diagnosis include polymerase chain reaction (PCR) and its variants, loop-mediated isothermal amplification (LAMP), nucleic acid hybridization, nucleic acid amplicon sequencing and phylogenetics, CRISPR-Cas-based methods, metagenomics, etc.

2.3.1 Loop-mediated isothermal amplification (LAMP)

Loop-mediated isothermal amplification is a rapid disease diagnosis tool becoming increasingly popular in plant pathology due to its high sensitivity, cost-effectiveness, and ability to visually verify positive results. A LAMP reaction consists of three steps, namely the initial step, the cycling amplification step, and the elongation step. In this method, six unique sequences present in the targeted nucleic acid are identified by a set of two internal primers, one backward loop primer, and a set of two outer primers. In tea, LAMP methodology has been developed for the detection of the blister blight pathogen *E. vexans* [12] and on-site diagnosis of the anthracnose pathogen *C. siamense* [13].

2.3.2 Polymerase chain reaction

Polymerase chain reaction has revolutionized plant disease diagnosis and pathogen detection by offering a highly sensitive and specific tool for identifying the presence of pathogens in plants. PCR operates by amplifying specific segments of the pathogen's nucleic acid through a series of temperature cycles, including denaturation, annealing, extension, and final extension. By utilizing primers that are specific to the target pathogen's genetic material, PCR allows for the identification of pathogens. Conventional end-point PCR, quantitative PCR (qPCR), multiplex PCR, and nested PCR are some variations of PCR methods that are widely used in plant pathology. PCR products are subjected to sequencing and compared with the existing data in the NCBI GenBank. This comparison process allows for the development of species-specific primers as needed, ensuring a more precise and tailored approach for pathogen identification.

The sequence information for most tea pathogens is limited. Hence, universal primers are used in most studies investigating the molecular identification and diversity of tea pathogens. Using the 28SrRNA sequence of *E. vexans* available in the NCBI GenBank, Chaliha et al. [10] developed a species-specific primer pair for the selective amplification of *E. vexans* (EVLSUF-EVLSUR). However, the combination of universal forward primer ITS1F and species-specific reverse primer EVLSUR allows for the identification of multiple strains, resulting in a larger PCR product.

An RT-qPCR method has been developed to quantify the *C. camelliae* (tea brown blight or tea anthracnose disease) fungal biomass in tea leaves. This method uses species-specific primers (**Table 2**) and tea plant-specific primers (forward primer: 5′-GACTCCGCTGGCACCTTAT-3′; reverse primer: 5′-GCCCTTCCGTCAATTCCT-3′) for normalization [14]. Compared to the traditional lesion area measurement approach, the proposed RT-qPCR method is highly sensitive and rapid. Further, this method allows for the differentiation of small differences in disease severity levels that may be undetectable by lesion size alone.

Pathogen [Reference]	Method	Name	Primer sequences
Exobasidium vexans [10, 12]	LAMP	Inner primers (EV_FIP, EV_BIP)	F-TACGGGGCTCTCACCCTCTATG + TGTAGAGGATGCTTTTGGCG R-GACCACCGAGCCTCTGTAAAGC + AGAAGACGTACACCTCCCAT
		Loop primers (EV_LF, EV_LB)	F-TTCCAGGGAACTCGGAAGGCA R-TCCTTCGACGAGTCGAGTAGTTTGG
		Outer primers (EV_F3, EV_B3)	F-TCTCTGGCCGGTATTTAGCT R-TGAAATCTGGCCGCAAGG
	End-point PCR	EVLSUF-EVLSUR	F-GCCGGATGAGCTCAAATTTG R-CCAAAACTGATGTTGGCCTG
		ITS1F-EVLSUR	F-CTTGGTCATTTAGAGGAAGTAA R-CCAAAACTGATGTTGGCCTG
Colletotrichum siamense [13]	LAMP	Inner primers (FIP, BIP)	F-GAAAAAGCACCGTCGTAGCCA+ TGAAGGATACCGACTCCGAA R-ATATGTTGACTCTGAATAGG+ ATGACGTGACGCAACTCG
		Loop primers (F-loop, B-loop)	F-AGCCTCGCGAATCTCCTC R-TCTTTGACCGCGACAACAAC
		Outer primers (F3, B3)	F-ACTATGATGGCCCGCAAGA R-GCTTCTCACCAATGGAGGTC
	End-point PCR	P-Cs	F-TGCTTTTTCGCTGCGATAT R-CCATGACGCATACCAATCAA
C. camelliae [14]	RT – qPCR	S572/S573	F-CCCGCATCTGGTAGACAAGA R-TGATAGCATGTGTCCCTCCG
		S576/S577	F-AAAGGTAGTGGCGGACCCT R-CCCAGTGCGAGACGTAAAGT
Tea Plant Necrotic Ring Blotch Virus [15]	RT – qPCR	qTPNRBV1	F-TGTGCGACGGACTCATTCTC R-TCAAGCATGCGAGTTTCCCT
		qTPNRBV2	F-CCCGTCATCGATACGCTCAA R-TACCTGGGGGAAGGTCTGAA
		qTPNRBV3	F-GATCGTCCTTCCCGATCTGC R-GTGCAGGCGTTCGCTTATTG
		qTPNRBV4	F-TTGGTTCGCTGTCCACCATT R-GAGCTGAAGCTGACGCATGA
Tea plant necrotic ring blotch virus [16]	RT-PCR	TPNRBV1	F-GCCCTGACAACGCAAAAGAACTGATG R-GTGACGGGATATTTTTGGACGACTGT
		TPNRBV2	F-GGGCCGGGGTGTGGAAAAACTT R-TTCTTATCATCCCGGCAAAACACA
		TPNRBV3	F-TTCGCCACTCACAAAGACAACAAACT R-GTAGCGGAGCGGAAAGAAAAGACT
		TPNRBV4	F-TCAGTGGCGCGATTATCAGAAGGTA R-CGCGCAAGAAGTCGGTCAAAAC

Pathogen [Reference]	Method	Name	Primer sequences
[17]	RT-PCR	TPNRBV	F-CCTTATGTCGACAGTTGCTAC R-CTAAGTCATCCATATGTGTGG
Tea plant line pattern virus [16]	RT-PCR	TPLPV1	F-AAGGTGGCGAGGTCAGTTTCAGTTCA R-TCCCCATAGGTTCATCTTGTAGCAGTCG
		TPLPV2	F-CCTATGGAGCTCTATGACGCAAATGAT R-TCGACAGAGAAGTGATGGGGAAATAC
		TPLPV3	F-TCAGATGCGACCAATGGAACTCAACT R-CCCGCGCTAATCCTCTCAAGACTAACTA

F, forward primer, R, reverse primer.

Table 2.
Species-specific primers for polymerase chain reaction (PCR) and loop-mediated isothermal amplification (LAMP)-based diagnosis of fungal and viral disease in tea.

Species-specific internal transcribed spacer (ITS) markers have been developed for stem and branch canker pathogens *B. dothidea* and *L. theobromae*, allowing detection through conventional PCR methods. This method is useful for identifying the latent infection of the pathogen in planting material and targeting species-specific fungicide treatment in infected tea plants [8].

Viral diseases are uncommon in tea. Recently, Tea Plant Necrotic Ring Blotch Virus (TRNBV) and Tea Plant Line Pattern Virus (TRLPV) have been identified in China. The genome of TRNRBV has four positive-sense single-stranded RNA, and TRLPV has three segments of single-strand RNA [16, 18]. Conventional and a SYBR Green RT-qPCR method have been developed for the detection of four different segments of TPNRBV RNA [15, 16]. Importantly, the RT-qPCR method offers a significant advantage: it is 100-fold more sensitive than conventional detection methods. **Table 2** provides details of specific primers for PCR and LAMP-based diagnosis of diseases in tea.

2.3.3 Metagenomics

Metagenomic sequencing is a cutting-edge technique in which the genetic material of entire communities of microorganisms in a particular sample is evaluated. This method delves into the hidden and complex world of microorganisms by directly sequencing and analyzing DNA or RNA extracted from a particular sample. In plant disease diagnosis, metagenomics provides an unbiased, comprehensive approach to identifying pathogens associated with the disease symptoms. This technique has the added advantage of identifying unculturable, unexpected emerging pathogens often missed by standard approaches. The first application of the metagenomics approach to pathogen identification in tea was performed in China. In this study, researchers identified seven putative viruses including two novel species, TPNRBV, and TPLPV from tea leaves exhibiting albino and chlorina symptoms [16].

2.3.4 Nucleic acid amplicon sequencing and phylogenetics

Nucleic acid sequencing and phylogenetic analysis have revolutionized plant disease diagnosis. These tools allow for accurate pathogen identification, disease origin tracking, and a deeper understanding of the evolutionary relationships between different pathogen strains. By sequencing specific genetic markers (e.g., 16S ribosomal RNA (16S rRNA) for bacteria and internal transcribed spacer (ITS) within ribosomal RNA for fungi) and comparing them against a reference database, pathogen/s can be precisely identified. While 16S rRNA and ITS are widely used, they often lack the resolution needed to distinguish closely related strains. In these cases, sequencing additional marker genes and follow-up phylogenetic analysis significantly increases taxonomic accuracy.

DNA sequencing and phylogenetic analysis have significantly advanced tea pathology, revealing unknown pathogens and correcting misidentifications. For instance, increasing leaf blight reports from India and China show symptom similarity, leading to misdiagnosis. DNA sequencing revealed at least three distinct pathogens, *Muyocopron laterale* [3], *Alternaria alternata* [19], and *Arthrinium arundinis* [20]. Similarly, diverse pathogens were identified for leaf spot diseases across China (**Table 1**) [4, 5, 21–23].

Macrophoma theicola was long thought to cause stem and branch canker. However, recent studies utilizing DNA sequencing and phylogenetic analysis have refined this identification to three species within Botryosphaeriaceae family [8]. Controversy surrounds the identity of the pathogen causing tea white scab disease, with some suggesting *Phyllosticta theaefolia* or *Elsinoe leucospila*. However, a comprehensive approach using DNA sequencing, morphological characterization, and pathogenicity tests identified *E. leucospila* as the prominent pathogen [24].

Multigene phylogenetic analysis has revealed that gray blight disease in tea is caused by a diverse group of species belonging to at least three genera: *Pseudopestalotiopsis*, *Neopestalotiopsis*, and *Pestalotiopsis* [25–32]. According to the earliest known report by Massee from Sri Lanka, brown blight disease in tea is caused by *C. camelliae*. However, recent multi-locus phylogenetic approaches have revealed at least 17 *Colletotrichum* species can cause the same disease with varying degrees of severity [33–38].

2.3.5 Whole genome sequencing

Whole genome sequencing (WGS) provides a comprehensive analysis of a pathogen's genetic blueprint, providing in-depth insights into its biology, virulence factors, and evolution. Unlike PCR-based methods and amplicon sequencing, whole genome sequencing can be conducted without the need for specific primers. In tea, genome sequencing of only very few pathogens is available. However, there is no evidence to show its application in pathogen identification yet. Kong et al. [38] sequenced the whole genome (67.74 Mb) of tea brown blight pathogen *C. camelliae* using the PacBio RSII sequencing platform and identified the pathogenicity-related gene clusters. The genome of tea leaf spot pathogen *D. segeticola* (33.4 Mb) has been sequenced, and its importance in studying pathogenicity determinants, pathogen evolution, and disease control strategies has been highlighted [15]. The whole genome of *Ps. theae* (50.41 Mb) has been sequenced and deposited at GenBank under the accession number JACWGD000000000 [39]. Whole genome sequences of two different strains of tea plant virus TPNRBV (Chinese isolate and Japanese isolate) are available [16, 18]. This information proved valuable to the design of species-specific primers for PCR-based diagnosis of viral diseases in tea.

2.4 Remote sensing and image analysis methods

Remote sensing and image analysis have emerged as powerful tools in the field of plant disease diagnosis, revolutionizing the way we detect, monitor, and manage diseases in agricultural and natural ecosystems. Remote sensing involves the use of specialized sensors attached to satellites, drones, or ground-based instruments, to capture data from a distance. These sensors collect a wide range of data, including optical, thermal, and hyperspectral images. Infrared thermal imaging, red, green, and blue (RGB) imaging, multispectral imaging, hyperspectral imaging, chlorophyll fluorescence imaging, 3D sensors, etc., are some examples of remote sensing and imaging technologies widely used in plant disease diagnosis.

2.4.1 Multispectral and hyperspectral imaging

Spectral imaging provides a marker for a plant's health by using single or multiple wavelengths in the electromagnetic spectrum. Healthy plants with their higher chlorophyll content absorb more red light and therefore reflect a higher proportion of near-infrared (NIR) than diseased plants. A diseased plant does the opposite. The information is combined in the form of different vegetation indices to estimate the plant's fitness or performance.

Hyperspectral imaging technology has been employed to identify anthracnose disease in tea [40, 41]. Compared to healthy leaves, diseased leaves exhibited reduced spectral reflectance near 550 nm and 770 nm, with an increase observed around 680 nm. The accuracy of the method was 94.12–94.28% [40].

Hyperspectral signatures in the wavelength range of 700 to 1000 nm (NIR region) differentiated blister blight-infected leaves from healthy ones [42]. Blister blight disease progresses through the translucent spot, mature blister, and necrotic stages. The reflectance decreased by 2.4, 4.5, and 30.3% as the disease progressed from the translucent spot to the necrotic stage, respectively. Spectral wavelengths at 560, 640, and 780 nm differentiated anthracnose, brown leaf spot disease, and tea white star diseased leaves from healthy leaves [43]. A hyperspectral image classification model CARS-LSTM (The competitive adaptive reweighted sampling-Long short-term memory) achieved 95% accuracy in identifying tea coal disease (*Neocapnodium theae* Hara), outperforming the RGB model, WT-ResNet18 [44].

2.4.2 RGB imaging

RGB imaging in plant disease diagnosis involves capturing digital images of plants using standard cameras, including smartphones and specialized imaging devices, and follow-up analysis and comparisons of its three features: color, texture, and shape. Artificial Intelligence (AI) and Machine Learning (ML) play pivotal roles in image analysis, revolutionizing how we interpret and extract meaningful information from visual data.

In the recent past, there has been a noticeable rapid increase in the use of RGB image analysis for tea disease diagnosis. The targets are foliar diseases, including blister blight, gray blight, brown blight/tea anthracnose, tea coal disease, algal leaf disease, leaf spot diseases, etc. Importantly, multiple diseases together with pest-infested leaves, were utilized to improve the differentiation potential and classification accuracy. Further, the above approach has laid the foundation for diagnosing multiple diseases and pest attacks on the same tea leaf. Among ML algorithms, Convolutional Neural Networks (CNNs) stand out as a popular choice, primarily attributed to their exceptional capabilities in image classification (**Table 3**).

Disease monitoring is also achieved with imaging and ML techniques. Analysis of RGB images using YOLOv8 and Residual Network 50 (Resnet50) CNN models have been effective for diagnosing and assessing the severity of blister blight disease [61]. Edwin Raj et al. [62] used RGB images to develop a methodology to spatially assess the percentage of disease with 95% accuracy. Furthermore, the method facilitates the

Diseases	Classification algorithm/architecture	Average accuracy	Ref.
Gray blight	Deep CNN based on MobileNet and VGG19 Net	98.99%	[45]
Gray and brown blights, algal, and red spots, *Helopeltis* pest attack	Deep CNN	96.56%	[46]
Red rust, red spider, thrips, and *Helopeltis* pest attack	Multi-class support vector machines	83%	[47]
Red, white, and algal leaf spots, bird's-eye spots, gray and brown blights	Deep CNN model (LeafNet)	90.16%	[48]
Algal leaf disease	Ensemble neural network	91%	[49]
Brown blight and algal spots	Support vector machine classifier	90%	[50]
Leaf blight diseases	Faster region-based convolutional neural networks (Faster R-CNN), and VGG16	84%	[51]
Tea red scab, tea red leaf spot, and tea leaf blight	Conditional deep convolutional generative adversarial networks (C-DCGAN), VGG16	90%	[52]
Tea red leaf spot and tea red scab	Gradient Boosting Decision Tree (GBDT)	95%	[53]
Leaf and bud blight and red scab	Custom DCNN	92.5%	[54]
Leaf lesions	Faster R-CNN and cascade R-CNN (CRCNN)	74–76.6%	[55]
Blister blight	Deep hashing with integrated autoencoders (DHIA) consisting of basic model (BM) and autoencoders (AE)	>97%	[56]
Algal leaf spot, gray and brown blight, white spot, red scab, and bud blight	CNN	94.45%	[57]
Algal leaf spot, anthracnose, and bird's-eye spot	CNN	95%	[58]
Algal leaf spot, leaf and bud blight, white scab	AX-RetinaNet	93.83%	[59]
Leaf blights	Lightweight and efficient LC3Net model with integration of a channel attention module (CAM)	92.29%	[60]

Table 3.
RGB imaging techniques and machine learning algorithms are used for artificial intelligence-based diagnosis of tea diseases.

mapping of disease occurrences at the estate level using smartphone GPS data and Geographic Information System (GIS).

Dahanayake et al. [63] developed an Android app for diagnosing tea leaf diseases using smartphone images. This app with Artificial Neural Networks (ANNs) allows farmers to diagnose diseases on-site, saving time and money. A study developed a CNN model and deployed it to the Platform-as-a-Service cloud for classifying tea leaf disease using smartphone images. Hence, the digital images captured using smartphones can be uploaded to predict the disease automatically [64].

3. Challenges in tea disease diagnosis

Tea is grown as an intensive monoculture crop. This system is vulnerable to the rapid spread of pests and diseases, making the problem unmanageable within a short time. Early detection and accurate identification of diseases and their causal agents are paramount for efficient and economically viable disease management. With the advancement of science, a combination of traditional and modern techniques is used in the diagnosis of tea diseases. However, there are several challenges to be faced. This section analyses the major challenges faced in tea disease diagnosis.

3.1 Disease symptoms with a wide range of characteristics

Some diseases may produce easily recognizable symptoms such as discoloration, wilting, or visible lesions, while others may exhibit subtle or less distinct symptoms that are easily overlooked. For instance, blister blight disease develops through different stages over 2 to 4 weeks. Though the leaf and stem blisters are easily identified and captured by various techniques, the first visible tiny to small translucent spots generally go unnoticed and are difficult to sense by imaging techniques. This makes early detection of the disease challenging even with modern imaging techniques.

3.2 Similar symptoms produced by different diseases/factors

Leaf blights caused by different pathogens show symptoms that overlap (Section 2.1). For example, brown blight and gray blight share common features in their symptom expression. Delimitation of these two diseases only by symptomology is always impossible. Under such conditions, culturing techniques or DNA-based techniques help with accurate identification. Leaf spot diseases also share common phenotypical expressions. Nutrient deficiency symptoms resemble leaf blights and leaf spots.

Sudden death of a branch or whole plant with attached leaves occurs when tea plants are infected with root pathogens. Though root symptoms may be different, the above-ground symptoms are common for most of the root diseases. Death of plants due to drought, white grub infestation, and root diseases also share similar features and always demand careful diagnosis for differentiation.

3.3 Simultaneous occurrence of more than one disease

Under certain environmental conditions, co-infections of fungal pathogens are common. Brown blight infection generally follows blister blight disease. The occurrence of brown blight and gray blights on the same leaf or plant is a common feature. The application of imaging techniques or DNA-based techniques is difficult under such conditions.

3.4 Tea canopy architecture and leaf orientation

Naturally, the tea plant grows as a tree in the wild. Periodic pruning and frequent harvesting maintain the plant as a bush. Close planting (12,500 plants per hectare in Sri Lanka) helps to form a continuous harvesting table. There are several generations (different age groups) of tea shoots at a time on the tea canopy. The overlapping shoot arrangement and leaf orientation always challenge image-based techniques in disease diagnosis.

Similarly, stem diseases such as cankers, wood rots, and stem blights cannot be diagnosed early with imaging techniques or monitoring at a distance. Careful observation of stems and branches is necessary to identify such conditions, as they may only become visible during pruning operations.

3.5 Asymptomatic infections

Several species of *Botryosphaeriaceae*, *Pestolotiopsis*, *Phomopsis* (*Diaphorthe*), etc., occur as endophytes or latent infections in tea plants where the pathogen(s) are present in the plant but do not cause visible symptoms. This can lead to undetected infections and disease spread.

3.6 Environmental factors, plant nutrient status, and plant factors

Environmental conditions, such as weather, humidity, soil quality, and plant nutrient status, can influence the development and appearance of disease symptoms. Different tea plant varieties respond to certain diseases differently, which can complicate diagnosis because the same pathogen may have different effects on different varieties. These variabilities can make it challenging to establish a definitive diagnosis protocol.

3.7 Limited diagnostic tools and expertise

The global tea industry slowly and gradually grabs new and modern technologies as tea is mostly produced in relatively low-income countries [65]. There is limited access to advanced diagnostic tools, such as imaging and image analysis, DNA-based tests, or specialized equipment and trained personnel, which are essential for accurate disease diagnosis in most of the tea-producing countries.

3.8 Climate change

Tea is produced in narrowly defined agroecological conditions as a rain-fed monocrop. The changes in the rainfall patterns, rising temperature, and increasing events of floods and drought events would lead to the emergence of new pathogens. The possible emergence of new diseases and changes in existing pathogen populations would make disease diagnosis challenging, demanding more sophisticated methods and techniques.

4. Way forward

It is forecasted that the global tea industry will grow with a compound annual growth rate (CAGR) of 6.7 percent from 2022 to 2031. Minimizing crop loss due to pests and diseases while ensuring pesticide-free quality products is a challenge to the growing tea industry under current climate change. There are several opportunities for improving tea disease diagnosis, which can lead to more effective disease management and increased crop yield in the tea industry. These opportunities encompass technological advancements, research, and collaboration.

As many tea-growing countries share common disease problems, global collaboration in disease diagnosis and studies on pathogens would help to bring appropriate technology rapidly at a shared cost. Once such technology is developed with coordinated effort, refinement at local conditions can be carried out. This collaborative approach could minimize duplication of work among tea-growing countries on the same disease problem. The traditional method of disease diagnosis with culturing and disease symptomology has been supported by DNA-based methods in the recent past for many tea diseases. The available information can be used by tea pathologists with modification if required.

The establishment of disease surveillance networks that involve collaboration among tea growers, researchers, and government agencies, along with timely sharing of disease data, can facilitate rapid response and containment. Training and educational programs for tea growers and plantation workers to enhance their knowledge of disease symptoms, prevention, and early detection techniques could help manage resource limitations.

Whole genome sequences are available only for brown blight pathogen *C. camelliae*, tea leaf spot pathogen *D. segeticola, P. theae,* and two different strains of tea plant virus TPNRBV despite more than 300 fungal pathogens being reported in tea. With global collaboration and wise utilization of resources, it is imperative to generate whole genome sequences of major tea pathogens. This information would help develop and implement advanced diagnostic tools, as well as high-throughput assays. These technologies can provide rapid and highly accurate identification of tea pathogens.

Although imaging and remote sensing technology are developing rapidly, most of the works are from academic research limiting agricultural applications. Hyperspectral imaging may be promising due to its high specificity and rapid data analysis. However, the use of drones or unmanned aerial vehicles for collecting data would remain challenging in situations where tea is grown under shade trees, particularly in India and Sri Lanka. To get the maximum potential of machine learning techniques, a large, verified dataset of images of diseased and healthy plants is essential. The establishment of quality data sets may overcome the problems faced by data scientists. Smartphone and mobile phone-based diagnostic tools will help share knowledge and improve the diagnostic capabilities of even the tea growers.

There are unexplored horizons in tea disease diagnostic methods. Nanotechnology-based pathogen detection, biosensors based on highly selective bio-recognition elements, and CRISPR-Cas-based detection systems have so far not experimented with tea pathogens. Although the application of ELISA for tea disease detection has not been reported much, the fabrication of test strips based on ELISA will be an innovative strategy for the detection of tea viruses in the future due to its ease of handling. Lateral flow assays would favor in-field quick diagnosis. It would be advantageous to develop multiple pathogen detection systems rather than the detection of a single pathogen.

5. Conclusions

An early and precise diagnosis of diseases is crucial, enabling prompt intervention and precise disease management plans. The global tea industry is gradually embracing modern methodologies rather than relying solely on symptomatic and culture-based methods. DNA and RNA-based diagnosis, genome sequencing, remote sensing, machine learning, and image processing technology to achieve automatic detection are widely used in tea. The adoption of current techniques and their field-level applications are, however, influenced by several factors. The future needs are discussed in detail.

DOI: http://dx.doi.org/10.5772/intechopen.1004903

Conflict of interest

The authors declare no conflict of interest.

Author details

Ganga Devi Sinniah[1*] and Niranjan Mahadevan[1,2]

1 Plant Pathology Division, Tea Research Institute of Sri Lanka, Talawakelle, Sri Lanka

2 Graduate School of Environmental and Life Sciences, Okayama University, Okayama, Japan

*Address all correspondence to: gdsinniah@gmail.com

References

[1] Pandey AK, Sinniah GD, Babu A, Tanti A. How the global tea industry copes with fungal diseases - Challenges and opportunities. Plant Disease. 2021;**105**(7):1868-1879. DOI: 10.1094/PDIS-09-20-1945-FE

[2] Chen ZM, Chen XF. An analysis of world tea pest fauna. Journal of Tea Science. 1989;**9**:13-22

[3] Jiang H, Zhang M, Zhou Y, Li X, Li J, Song C. First report of *Muyocopron laterale* causing a new leaf disease of *Camellia sinensis* in China. Beverage Plant Research. 2022;**2**:1-4. DOI: 10.48130/bpr-2022-0010

[4] Chen YJ, Wan YH, Zou LJ, Tong HR. First report of leaf spot disease caused by *Epicoccum layuense* on *Camellia sinensis* in Chongqing, China. Plant Disease. 2020;**104**(7):2029. DOI: 10.1094/PDIS-09-19-1906-PDN

[5] Deng X, Yang J, Wan Y, Han Y, Tong H, Chen Y. Characteristics of leaf spot disease caused by *Didymella* species and the influence of infection on tea quality. Phytopathology. 2023;**113**:516-527. DOI: 10.1094/PHYTO-06-22-0202-R

[6] Yin QX, An XL, Wu X, Dharmasena DSP, Li DX, Jiang SL, et al. First report of *Alternaria longipes* causing leaf spot on tea in China. Plant Disease. 2021;**105**(12):4167. DOI: 10.1094/PDIS-07-20-1583-PDN

[7] Yin QX, Jiang SL, Li DX, Huang HL, Wang Y, Wang DL, et al. First report of *Epicoccum nigrum* causing brown leaf spot in tea in Guizhou Province, China. Plant Disease. 2022;**106**(1):321. DOI: 10.1094/PDIS-04-21-0815-PDN

[8] Pradeepa NHL, Weerasena OVDSJ, Liyanaarachchi CJ, Karunajeewa DGNP, Mahindapala KGJP, Mahinda TGN, et al. Morphological and molecular characterization of fungi causing canker of tea (*Camellia sinensis*). In: Proceedings of Fifth Plantation Crop Symposium, 15-17 October 2014. Sri Lanka: BMICH, Colombo; 2014

[9] Sinniah GD, Munasinghe CE, Mahadevan N, Jayasinghe SK, Kulatunga DCM. Recent incidence of collar canker and dieback of tea (*Camellia sinensis*) caused by *Fusarium solani* species complex in Sri Lanka. Australasian Plant Disease Notes. 2017;**12**(41):1-6. DOI: 10.1007/s13314-017-0262-5

[10] Chaliha C, Kaladhar VC, Doley R, Verma PK, Kumar A, Kalita E. Bipartite molecular approach for species delimitation and resolving cryptic speciation of *Exobasidium vexans* within the *Exobasidium* genus. Computational Biology and Chemistry. 2021;**92**:107496. DOI: 10.1016/j.compbiolchem.2021.107496

[11] Zhao XZ, Wang Y, Ren YF, Li DX, Chen Z. The morphology observation of infection process for the pathogen *Exobasidium vexans* of tea blister blight against tea leaf. Chinese Agricultural Science Bulletin. 2020;**34**:117-122. DOI: 10.21512/commit.v13i2.5886

[12] Chaliha C, Srivastava R, Kalita E, Sahoo L, Verma PK. Rapid and precise detection of cryptic tea pathogen Exobasidium vexans: RealAmp validation of LAMP approach. World Journal of Microbial Biotechnology. 2022;**39**(2):52. DOI: 10.1007/s11274-022-03506-y

[13] Zou H, Li T, Zhang J, Shao H, Kageyama K, Feng W. Rapid detection of *Colletotrichum siamense* from

infected tea plants using filter-disc DNA extraction and loop-mediated isothermal amplification. Plant Disease. 2023;**108**(1):35-40. DOI: 10.1094/PDIS-05-23-0913-SC

[14] He S, Chen H, Wei Y, An T, Liu S. Development of a DNA-based real-time PCR assay for the quantification of *Colletotrichum camelliae* growth in tea (*Camellia sinensis*). Plant Methods. 2020;**16**(17):1-11. DOI: 10.1186/s13007-020-00564-x

[15] Ren H, Chen Y, Zhao F, Ding C, Zhang K, Wang L, et al. Quantitative distribution and transmission of tea plant necrotic ring blotch virus in *Camellia sinensis*. Forests. 2022;**13**(8):1306. DOI: 10.3390/f13081306

[16] Hao X, Zhang W, Zhao F, Liu Y, Qian W, Wang Y, et al. Discovery of plant viruses from tea plant (*Camellia sinensis* (L.) O. Kuntze) by metagenomic sequencing. Frontiers in Microbiology. 2018;**9**:2175. DOI: 10.3389/fmicb.2018.02175

[17] Chen X, Shen J, Li M, Gao Y, Du Z, Gao F. The occurrence and genetic variability of tea plant necrotic ring blotch virus in Fujian Province, China. Forests. 2023;**14**:1755. DOI: 10.3390/f14091755

[18] Maruyama N, Iwabuchi N, Nishikawa M, Nijo T, Yoshida T, Kitazawa Y, et al. Complete genome sequence of tea plant necrotic ring blotch virus detected from a tea Plant in Japan. Microbiology Resource Announcements. 2022;**11**(6):e00323-22. DOI: 10.1128/mra.00323-22

[19] Chen YJ, Meng Q, Zeng L, Tong HR. Phylogenetic and morphological characteristics of *Alternaria alternata* causing leaf spot disease on *Camellia sinensis* in China. Australasian Plant Pathology. 2018;**47**:335-342. DOI: 10.1007/s13313-018-0561-0

[20] Thangaraj K, Cheng LL, Deng C, Deng WW, Zhang ZZ. First report of leaf blight caused by *Arthrinium arundinis* on tea plants in China. Plant Disease. 2019;**103**(12):3282. DOI: 10.1094/PDIS-06-19-1324-PDN

[21] Bao XT, Dharmasena DSP, Li DX, Wang X, Jiang SL, Ren YF, et al. First report of *Epicoccum sorghinum* causing leaf spot on tea in China. Plant Disease. 2019;**103**(12):3282. DOI: 10.1094/PDIS-06-19-1296-PDN

[22] Yin QX, Jiang SL, Li DX, Huang HL, Wang Y, Wang DL, et al. First report of *Epicoccum nigrum* causing brown leaf spot in tea in Guizhou Province, China. Plant Disease. 2022;**106**(1):321. DOI: 10.1094/PDIS-04-21-0815-PDN

[23] Li DX, Bao XT, Ren YF, Song BA, Chen Z, Wang Y. First report of *Lasiodiplodia theobromae* causing leaf spot on tea plant in Guizhou Province of China. Plant Disease. 2018;**103**:374. DOI: 10.1094/PDIS-06-18-1032-PDN

[24] Lingyun Z, Yunfeng L, Chunyan J, Hongyan L, Wei L, Fen X, et al. Identification of the pathogen responsible for tea white scab disease. Journal of Phytopathology. 2020;**168**:28-35. DOI: 10.1111/jph.12862

[25] Chen Y, Zeng L, Shu N, Jiang M, Wang H, Huang Y, et al. *Pestalotiopsis*-like species causing gray blight disease on *camellia sinensis* in China. Plant Disease. 2018;**102**:98-106. DOI: 10.1094/PDIS-05-17-0642-RE

[26] Maharachchikumbura SSN, Chukeatirote E, Guo LD, Crous PW, McKenzie EHC, Hyde KD. *Pestalotiopsis* species associated with *Camellia sinensis* (tea). Mycotaxon. 2013;**123**:47-61. DOI: 10.5248/123.47

[27] Chen Y, Wan Y, Zeng L, Meng Q, Yuan L, Tong H. Characterization of *Pestalotiopsis chamaeropis* causing gray blight disease on tea leaves (*Camellia sinensis*) in Chongqing, China. Canadian Journal of Plant Pathology. 2021;**43**:413-420. DOI: 10.1080/07060661.2020.1816582

[28] Pandey AK, Hubbali M, Vandana Dutta P, Babu A. Characterization and identification of fungicide insensitive Pestalotiopsis-like species pathogenic to tea crop in India. World Journal of Microbiology and Biotechnology. 2023;**39**(1):34. DOI: 10.1007/s11274-022-03474-3

[29] Tsai I, Chung CL, Lin SR, Hung TH, Shen TL, Hu CY, et al. Cryptic diversity, molecular systematics, and pathogenicity of genus *Pestalotiopsis* and allied genera causing gray blight disease of tea in Taiwan, with a description of a new *Pseudopestalotiopsis* species. Plant Disease. 2021;**105**:425-443. DOI: 10.1094/PDIS-05-20-1134-RE

[30] Chen L, Li H, Jiao W, Tao M, Lv C, Zhao M, et al. Genetic variation and demographic history analysis of *Pestalotiopsis*, *Pseudopestalotiopsis*, and *Neopestalotiopsis* fungi associated with tea (*Camellia sinensis*) inferred from the internal transcribed spacer region of the nuclear ribosomal DNA. Plant Pathology. 2021;**70**:699-711. DOI: 10.1111/ppa.13315

[31] Wang S, Mi X, Wu Z, Zhang L, Wei C. Characterization and pathogenicity of *Pestalotiopsis*-like species associated with gray blight disease on *Camellia sinensis* in Anhui Province, China. Plant Disease. 2019;**103**:2786-2797. DOI: 10.1094/PDIS-02-19-0412-RE

[32] Wang Y, Xiong F, Lu Q, Hao X, Zheng M, Wang L, et al. Diversity of *Pestalotiopsis*-like species causing gray blight disease of tea plants (*Camellia sinensis*) in China, including two novel *Pestalotiopsis* species, and analysis of their pathogenicity. Plant Disease. 2019;**103**:2548-2558. DOI: 10.1094/PDIS-02-19-0264-RE

[33] Chen Y, Qiao W, Zeng L, Shen D, Liu Z, Wang X, et al. Characterization, pathogenicity, and phylogenetic analyses of *Colletotrichum* species associated with brown blight disease on *Camellia sinensis* in China. Plant Disease. 2017;**101**:1022-1028. DOI: 10.1094/PDIS-12-16-1824-RE

[34] Liu F, Weir BS, Damm U, Crous PW, Wang Y, Liu B, et al. Unravelling *Colletotrichum* species associated with camellia: Employing ApMat and GS loci to resolve species in the *C. gloeosporioides* complex. Persoonia: Molecular Phylogeny and Evolution of Fungi. 2015;**35**:63-86. DOI: 10.3767/003158515X687597

[35] Wan Y, Zou L, Zeng L, Tong H, Chen Y. A new *Colletotrichum* species associated with brown blight disease on *Camellia sinensis*. Plant Disease. 2021;**105**(5):1474-1481. DOI: 10.1094/PDIS-09-20-1912-RE

[36] Wang YC, Hao XY, Wang L, Xiao B, Wang XC, Yang YJ. Diverse *Colletotrichum* species cause anthracnose of tea plants (*Camellia sinensis* (L.) O. Kuntze) in China. Scientific Reports. 2016;**6**:35287. DOI: 10.1038/srep35287

[37] Shi NN, Du YX, Ruan HC, Yang XJ, Dai YL, Gan L, et al. First report of *Colletotrichum fructicola* causing anthracnose on *Camellia sinensis* in Guangdong Province, China. Plant Disease. 2017;**102**:241

[38] Kong L, Chen J, Dong K, Shafik K, Xu W. Genomic analysis of *Colletotrichum camelliae* responsible for tea brown blight disease. BMC

Genomics. 2023;**24**:528. DOI: 10.1186/s12864-023-09598-6

[39] Zheng S, Chen R, Wang Zhe Liu J, Cai Y, Peng M, Zhang T, et al. High-quality genome assembly of *Pseudopestalotiopsis theae*, the pathogenic fungus causing tea gray blight. Plant Disease. 2021;**105**:3723-3726. DOI: 10.1094/PDIS-02-21-0318-A

[40] Zhao X, Zhang J, Huang Y, Tian Y, Yuan L. Detection and discrimination of disease and insect stress of tea plants using hyperspectral imaging combined with wavelet analysis. Computers and Electronics in Agriculture. 2022;**193**:106717. DOI: 10.1016/j.compag.2022.106717

[41] Yuan L, Yan P, Han W, Huang Y, Wang B, Zhang J, et al. Detection of anthracnose in tea plants based on hyperspectral imaging. Computers and Electronics in Agriculture. 2019;**167**:105039. DOI: 10.1016/j.compag.2019.105039

[42] Peiris CT, Marasinghe BS, Sinniah GD, Fonseka PU. Detection and estimation of tea blister using Spectrorediometeric analysis. In: Proceedings of the First National Symposium of Sri Lanka Association for Mycology and Plant Pathology: Plant Health. Kandy: Oak Ray Regency; 2019. p. 24 [Accessed: August 30th, 2019]

[43] Chen J, He Q. Tea disease spot recognition based on image feature points extraction and matching. Global Nest Journal. 2020;**22**:492-501. DOI: 10.30955/gnj.003375

[44] Xu Y, Mao Y, Li H, et al. A deep learning model for rapid classification of tea coal disease. Plant Methods. 2023;**19**:98. DOI: 10.1186/s13007-023-01074-2

[45] Pandian JA, Nisha SN, Kanchanadevi K, Pandey AK, Rima SK. Grey blight disease detection on tea leaves using improved deep convolutional neural network. Computational Intelligence and Neuroscience. 2023:1-11. Article ID: 7876302. DOI: 10.1155/2023/7876302

[46] Datta S, Gupta N. A novel approach for the detection of tea leaf disease using deep neural network. Procedia Computer Science. 2023;**218**:2273-2286. DOI: 10.1016/j.procs.2023.01.203

[47] Mukhopadhyay S, Paul M, Pal R, De D. Tea leaf disease detection using multi-objective image segmentation. Multimedia Tools and Applications. 2021;**80**:753-771. DOI: 10.1007/s11042-020-09567-1

[48] Chen J, Liu Q, Gao L. Visual tea leaf disease recognition using a convolutional neural network model. Symmetry (Basel). 2019;**11**(3):343. DOI: 10.3390/sym11030343

[49] Karmokar BC, Samawat Ullah M, Kibria Siddiquee M, Md. Rokibul Alam K. Tea leaf diseases recognition using neural network ensemble. International Journal of Computer Applications. 2015;**114**:27-30. DOI: 10.5120/20071-1993

[50] Hossain MS, Mou RM, Hasan MM, Chakraborty S, Abdur RM. Recognition and detection of tea leaf's diseases using support vector machine. In: Proceedings of 14th International Colloquium on Signal Processing and its Application. CSPA. Penang, Malaysia: IEEE- Institute of Electrical and Electronics Engineers; 2018. pp. 150-154. DOI: 10.1109/CSPA.2018.8368703

[51] Hu G, Wang H, Zhang Y, Wan M. Detection and severity analysis of tea leaf blight based on deep learning. Computers and Electrical Engineering.

2021;**90**:107023. DOI: 10.1016/j.compeleceng.2021.107023

[52] Hu G, Wu H, Zhang Y, Wan M. A low shot learning method for tea leaf's disease identification. Computers and Electronics in Agriculture. 2019;**163**:104852. DOI: 10.1016/j.compag.2019.104852

[53] Meng S, Wang S, Zhou T, Shen J. Identification of tea red leaf spot and tea red scab based on hybrid feature optimization. Journal of Physics Conference Series. 2020;**1486**:052023. DOI: 10.1088/1742-6596/1486/5/052023

[54] Prabu S, Bapu BRT, Sridhar S, Nagaraju V. Tea plant leaf disease identification using hybrid filter and support vector machine in classifier technique. In: Borah S, Mishra SK, Mishra BK, Balas VE, Polkowski Z, editors. Advances in Data Science and Management, Lecture Notes on Data Engineering and Communications Technologies. Singapore: Springer; 2022. p. 86. DOI: 10.1007/978-981-16-5685-9_58

[55] Chen XM, Lin CC, Lin SR, Chen SF. Application of region-based convolution neural network on tea diseases and harming insects identification. In: American Society of Agricultural and Biological Engineers (ASABE) Annual International Virtual Meeting 2021. St. Joseph, Michigan: American Society of Agricultural and Biological Engineers; 2021. p. 2100872. DOI: 10.13031/aim.202100872

[56] Jayapal SK, Poruran S. Enhanced disease identification model for tea plant using deep learning. Intelligent Automation and Soft Computing. 2023;**35**:1261-1275. DOI: 10.32604/iasc.2023.026564

[57] Latha RS, Sreekanth GR, Suganthe RC, Rajadevi R, Karthikeyan S, Kanivel S, et al. Automatic detection of tea leaf diseases using deep convolution neural network. In: International Conference on Computer Communication and Informatics, ICCCI 2021. Coimbatore, India: IEEE- Institute of Electrical and Electronics Engineers; 2021. DOI: 10.1109/ICCCI50826.2021.9402225

[58] Rosyidah EA, Hadi AF, Dewi YS. The classification of tea leaf disease using CNN image classifier. In: Proceedings of the 1st International Conference on Neural Networks and Machine Learning 2022 (ICONNSMAL 2022). The Netherlands, Dordrecht: Atlantis Press; 2022. pp. 89-112. DOI: 10.2991/978-94-6463-174-6_10

[59] Bao W, Fan T, Hu G, Liang D, Li H. Detection and identification of tea leaf diseases based on AX-RetinaNet. Science Report. 2022;**12**(1):2183. DOI: 10.1038/s41598-022-06181-z

[60] Jiang Y, Lu L, Wan M, et al. Detection method for tea leaf blight in natural scene images based on lightweight and efficient LC3Net model. Journal of Plant Diseases and Protection. 2023;**131**:209-225. DOI: 10.1007/s41348-023-00807-8

[61] Hewawitharana WGHAU, Nawarathne UMLA, Hassan ASF, Wijerathna LM, Sinniah GD, Vidhanaarachchi SP, et al. Effectiveness of using deep learning for blister blight identification in Sri Lankan tea. In: International Conference on Smart Computing and Systems Engineering, 2023. Sri Lanka: University of Kelaniya, IEEE- Institute of Electrical and Electronics Engineers; 2023 [Accessed: June 29th, 2023]

[62] Edwin, Raj E, Ashokraj S, Radhakrishnan B. Detection of blister blight disease in tea using smartphone and image processing technique.

Newsletter UPASI Tea Research Foundation. 2018;**28**(2):235-241

[63] Dahanayake VS, Samaraweera WJ, Kulasekara DMR. Smart tea leaves disease analyser: Mobile based disease detecting and solutions providing system. In: Proceedings of International Research Conference (KDU IRC). Ratmalana, Sri lanka: General Sir Kotelawala Defense University; 2018. Available from: http://ir.kdu.ac.lk/handle/345/2513

[64] Lanjewar MG, Panchbhai KG. Convolutional neural network based tea leaf disease prediction system on smart phone using paas cloud. Neural Computing and Applications. 2023;**35**:2755-2771

[65] FAO/EST. International tea market: Market situation, prospects and emerging issues. 2022

Section 2

Diseases Management

Chapter 5

Trichoderma spp.: Approach for Bio-Control Agent

Lovely Bharti, Kajol Yadav and Ashok Kumar Chaubey

Abstract

The novel technologies in all areas of agriculture have improved agricultural production, but some modern practices cause environmental pollution and human hazards. The recent challenge faced by advanced farming has been to achieve higher yields. Thus, there is an immediate need to find eco-friendly solutions. Among the various types of species being used as biocontrol agents, fungi of the genus *Trichoderma* are a very large group of microorganisms widely used as biocontrol agents against different kinds of plant pathogens. *Trichoderma* spp. are asexual, free-living organisms that are abundantly present in all types of agricultural soils. Recent studies have shown that *Trichoderma* can not only prevent diseases but also promote plant growth, improve nutrient utilization efficiency, enhance plant resistance, and improve the agrochemical pollution environment. *Trichoderma* spp. behaves as a low-cost, effective, and eco-friendly biocontrol agent for different crop species. This chapter provides information on *Trichoderma* as a biocontrol agent, its biocontrol activity, and plant disease management programs.

Keywords: *Trichoderma*, biological control, plant growth, agriculture, plant pathogens

1. Introduction

Plant diseases caused by viruses, bacteria, fungi, or other macroorganisms have a significant impact on the reduction in food production, growth, and development in agriculture, which results in serious economic losses every year. The first option that farmers frequently use to control plant diseases is chemical pesticides. The primary benefit of these pesticides is their immediate use and "solution" to the issue. Chemical pesticides used over an extended period of time can contaminate soil and water, affecting both humans and animals, while occasionally leaving toxic residues that can encourage the growth of particular resistant species. In addition, pesticides have negative impacts on soil microbiomes, soil-dwelling species (such as beneficial insects, like pollinators), and the overall health of terrestrial and aquatic ecosystems [1]. In recent decades, Chemical pesticides, which are the most widely used technique for defending plants against fungal diseases, have placed serious pressure on the agricultural environment [2].

In order to avoid the negative effects of chemical pesticides, researchers are searching for alternate solutions to these issues, such as the use of biocontrol agents (BCAs) for disease control, minimizing the damage caused by plant pathogens, and an integrated approach with other chemicals that is environmentally safe. One of these methods is the use of BCAs, which are based on living microorganisms or their metabolites and products of natural origin that reduce the population of plant pathogens [3]. These biological disease management strategies are effective, long-lasting, eco-friendly, affordable, and safe for human health. At present, integrated pest management strategies and avoidance or regulation of pesticides by using *Aspergillus, Gliocladium, Trichoderma, Ampelomyces, Candida*, and *Coniothyrium* are some examples of fungal agents, and other bacterial agents include *Pseudomas, Bacillus*, and *Agrobacterinum* [4]. Among these BCAs, *Trichoderma* spp. is one of the most adaptable and has been used for a long time to control plant pathogenic fungus and lessen the usage of pesticides on economically significant crops. *T. harzianum* and *T. virens* prevented the pathogenic fungus from the growth of *Ganoderma* [5].

Fungi that act as biocontrol agents have become useful tools in the modern agricultural system. They have the capacity to ameliorate abiotic stresses like drought, salinity, extremely high or low temperatures, and heavy metal impacts, as well as the harmful effects of plant pathogens. *Trichoderma* spp., which are beneficial fungi, have received a lot of attention because of their high reproductive capacity, ability to survive in adverse conditions, prolific production of secondary metabolites, and resistance to plant pathogenic fungi [6]. Additionally, they have been used in biotechnological applications and are crucial to agricultural endeavors because of their capacity to reduce biotic (species and vegetable variety and edaphic microbial interactions) and abiotic stresses (type of soil, water potential, temperature, and pH) and improve plant growth and yield [7]. Colonization with *Trichoderma* and plant root is the initial stage of successful interaction, signal exchange, and elicitor generation that results in symbiotic relationship between them. Produced enzymes and antibiotics move toward fungal pathogens, successfully breaking down their hyphae to allow entry into the host cell. Both enzymes and antibiotics have antifungal properties and work together to combat fungi [8]. *Trichoderma* are known to be resistant to *Alternaria alternate, Botrytis cinerea, Rhizoctonia solani, Sclerotinia sclerotiorum, Pythium* spp., and *Fusarium* spp. [9, 10], as well as nematode [11]. After successful colonization, they cause induced systemic resistance (ISR) in the plant, which results in the signaling of several hormones and eventually promotes development [8]. Abiotic factors like salt, drought, heavy metal buildup, and severe temperatures have an impact on crop yield all around the world. However, recent studies have shown that *Trichoderma* induces tolerance against abiotic stresses and improves growth in plants [12] like radish, cucumber, pepper, bottle gourd, periwinkle, bitter gourd, chrysanthemum, lettuce, and tomato [13]. *Trichoderma*-colonized plants produce substances like auxins (indole-3-acetic acid: IAA), ethylene (ET), gibberellins (GA), plant enzymes, antioxidants, phytoalexins, and phenols that offer long-lasting resistance to abiotic stresses and improve the root system's ability to branch. The main advantage of priming the plant for certain stress responses is that it provides a faster and stronger response if the stress occurs again.

The purpose of this chapter is to describe the role of *Trichoderma* species in detail as a biocontrol agent and biofertilizer and provide a brief overview of the commercially available products of *Trichoderma* species for application.

2. An overview of the genus *Trichoderma*

Trichoderma is a genus of filamentous, facultative, primarily asexual (the teleomorphic forms are Hypocrea), and anaerobic fungus that are widely distributed around the world and typically colonize decaying wood and other types of organic plant matter. *Trichoderma* is an essential component in the mycobiome of a wide range of soil ecosystems (such as farmland, prairie, forests, salt marshes, and deserts), including temperate and tropical areas, Antarctica, and tundra [2]. *Trichoderma* genus is a group of saprotrophic fungi that are widely distributed and frequently inhabit woody plants as endophytes [14]. The systematics and taxonomy of these fungi have changed when Persoon first coined the word "*Trichoderma*" in 1794 [15]. According to the present genus, it is a member of the eukaryota domain, fungi kingdom, ascomycota division, pezizomycotina subdivision, sordariomycetes class, hypocreales order, and hypocreaceae family. Over 370 *Trichoderma* species have previously been identified in the genus *Hypocrea/Trichoderma*, such as *T. harzianum, T. viride, T. asperellum, T. hamatum, T. atroviride, T. koningii, T. longibrachiatum,* and *T. aureoviride* [16, 17]. *Trichoderma* species are effective biocontrol agents in soil ecosystems because of their capacity for rapid growth, ability to use a variety of substrates, and resistance to numerous toxic chemicals, including fungicides (such as azoxystrobin, 3,4-dichloroaniline, and trifloxystrobin), herbicides, and other organic pollutants [18, 19].

Trichoderma species have the potential to act as biological plant protection agents. This was initially described in the early 1930s. According to the findings of researcher Weindling [20], the *T. lignorum* strain uses necrotrophic mycoparasitism to defend citrus seedlings from the pathogen *Rhizoctonia solani*. Since then, *Trichoderma's* biocontrol properties have been thoroughly studied for the treatment of diseases caused by several types of soil phytopathogens [21]. The mechanisms by which *Trichoderma* decreases the occurrence of plant diseases include competition for nutrients and space, the synthesis of antifungal metabolites, mycoparasitism, production of lytic enzymes that degrade the cell walls of fungal plant pathogens, as well as the induction of plant resistance [2]. According to Di Marco et al. [22], the *T. harzianum, T. hamatum, T. longibrachiatum, T. koningii, T. viride, T. polysporum*, and *T. asperellum* have the most effective biocontrol capabilities. Numerous studies have shown that the majority of *Trichoderma* spp. can create bioactive compounds and have antagonistic effects on plant pathogenic fungi [23]. These bioactive compounds, such as cell wall-degrading enzymes and secondary metabolites, can effectively increase crop resistance, lower plant illnesses, and stimulate growth of plants [24].

Trichoderma has a wide range of applications in agriculture due to its biocontrol, biostimulation, and biofertilization abilities. *Trichoderma* is recognized as the genus with the highest potential for biocontrol as it has the most isolated antifungal bioactive chemicals [25]. According to Rush et al. [25], *Trichoderma* species account for more than 60% of the fungal BCAs. For example, approximately 250 *Trichoderma*-derived biofungicides are used in India, but compared to biological management, Indian farmers still rely more heavily on synthetic chemical fungicides [26].

2.1 *Trichoderma* biology

The mycoflora of the genus *Trichoderma*, which are typically global in distribution and exhibit a wide range of genetic variety, are generally found in areas with access to decomposing plant materials, primarily cellulosic materials [27]. *Trichoderma* species are considered to be an imperfect fungus that belongs to the order Hypocreales of the Ascomycota. They can be easily separated from natural soil, decaying plant organic matter, and wood. *Trichoderma* multiplies and grows very quickly in different nutrient sources like Malt Agar (MA), Czapek Dox Agar (CDA), and Potato Dextrose Agar (PDA). It also produces conidia/spores of various shades that are characterized by a

Figure 1.
Three different isolated strains of Trichoderma *spp. [28].*

green color, especially at the center of a growing spot or in concentric ring-like zones on the agar surface (**Figure 1**), and some species also produce thick-walled chlamydospores. The most significant characteristic of this genus is its ability to parasitize other pathogenic mycoflora, particularly those that are associated with root rot and wilt diseases [29]. *Trichoderma* species have been described as endophytic fungi, even though they are typically found as opportunistic plant symbionts in all types of soil, including agricultural soil, orchard soil, and forest soil [30].

2.2 Morphological characteristics

Trichoderma species are mostly identified based on their morphology, although this method is not precise enough to distinguish between species' differences in diversity. According to [31], most *Trichoderma* cultures develop quickly in the ideal range of 25–30°C but not at 35°C, yet some species grow well at 35°C. This plays a key role in differentiating species with similar morphology. *T. harzianum* can be distinguished from species like *T. atroviride* and *T. aggressivum* that have a similar morphology by growing it at 35°C. After 96 hours, *T. harzianum* develops well and sporulates at 35°C, whereas *T. aggressivum* and *T. atroviride* cannot have colonies with a radius higher than 5 mm after 96 hours. Mycelia development and color features can be more easily identified in a rich media like Potato Dextrose Agar (PDA). *Trichoderma* uses a number of compounds, including carbon and nitrogen sources, for its sporulation. Sporulates of *Trichoderma* also produce powder masses with branched conidiophore bearing bright green, or occasionally colorless, grayish, or brownish, conidia [32], which is characteristic of genera such as *Myrothecium, Clonostachys*, and *Aspergillus* as well as *Penicillium*, respectively. Most of the time, their surfaces are smooth, but other species, such *T. viride*, have rough conidia [31]. The conidiophore is poorly defined, but it is mostly branched and contains unicellular conidia and phialides at the tip of the branched hyphal system that are not visible on one-week-old media. Phialides of *Trichoderma* spp. are often sterile hyphae creeping septate, forming a flat, firm, and tuft conidiophores erect originating from a short branch, whereas conidia have double-layered walls that have an electron-dense rough outer layer (epispore) and a moderately electron-dense inner layer. Conidia color morphology varies from species to species, but it is normally green or can be gray, white, and yellow. Colonies of *Trichoderma* spp. can develop slowly or swiftly, depending on the species. Their aerial mycelium is frequently restricted, floccose to arachnoid, and reverse colorless to dull

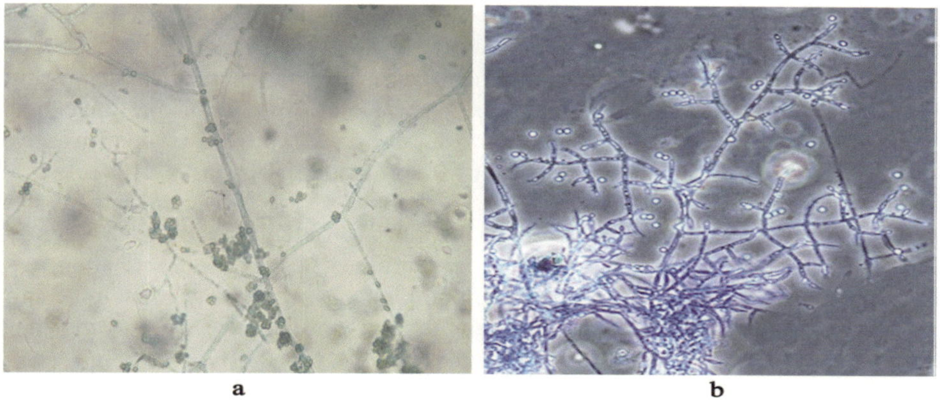

Figure 2.
Phialides and phialospores of (a) Trichoderma harzianum *and (b)* Trichoderma viride *(taken from the Université de Bretagne Occidentale website).*

yellow. Various isolates have a distinct aroma that is similar to coconut. Conidiation may be variously vibrant, loosely tufted, or produce compact pustules that initially appear white but eventually turn green (rarely brown). The conidiophores of *T. harzianum* with phialides and phialospores are depicted in **Figure 2**. Shah et al. [33] claim that the light green conidia of *T. viride* are globose to subglobose, whereas those of *T. harzianum* are globose. *T. pseudokoningii* showed little, light green conidia.

3. Biocontrol mechanisms of *Trichoderma* spp. against phytopathogens

Plant diseases are caused by the interaction among various components, including the host, pathogens, and environment, that is, the disease triangle. Bioagents are the organisms that prevent disease by interacting with different disease triangles. The interaction of pathogens and bioagents allows for modification of the soil environment to establish favorable conditions for effective biocontrol techniques against plant disease. Biocontrol agents involve a variety of processes in achieving disease control. However, conclusive evidences for the involvement of a specific factor in biological control are identified by the strong correlation between the factor's appearance and biological control. The study of various strategies for managing phytopathogens and plant diseases is the most significant and fascinating aspect of *Trichoderma*. *Trichoderma* uses competition, parasitism, antibiosis, induction of defense responses in host plants, and/or a combination of these strategies to control a variety of organisms.

3.1 Mycoparasitism of *Trichoderma*

A complex mechanism known as mycoparasitism or hyperparasitism allows an antagonistic fungus (mycoparasite) to parasitize on another fungus (host) and ultimately result in the death of pathogen cells [2]. The majority of the *Trichoderma* fungus is classified as a necrotrophic mycoparasite. *Trichoderma* parasitize a variety of mycoparasites, especially soilborne pathogens. Over 75 *Trichoderma* species have been identified with high mycoparasite potential and have the ability to lyse and attack plant pathogenic fungi like *Alternaria alternata, Botrytis cinerea, Rhizoctonia solani, Sclerotinia sclerotiorum, Pythium spp*., and *Fusarium spp*. [10]. *Trichoderma* necrotrophs have a mycoparasitic effect on fungi, which involves chemotaxis and prey sensing, attachment to the host, and physical attack through vigorous branching and coiling around the hyphae of the host. Moreover, *Trichoderma* can also create penetration structures that are similar to pathogen appressoria, or appressoria-like structures [34]. The act of mycoparasitism consists of three fundamental phases. This role can be performed in the rhizosphere of plants, an ecosystem where *Trichoderma* has successfully colonized and where the biological control of potential pathogens is crucial to prevent plant diseases. First, *Trichoderma* identifies its host (or potential plant pathogenic fungus), in which the formation of oligochitins has been considered as a sensor molecule [35]. Similar to this, it is understood that during the preceding stage, a number of gene-encoding proteases and oligopeptide transporters are expressed prior to interaction with the fungal host. Second, hydrophobin-like proteins might be useful when *Trichoderma* comes into contact with the plant pathogenic fungus, which results in the production of papillae or appressoria-like structures. The third step takes place when *Trichoderma* coils around the pathogen hyphae and begins to degrade it by producing cell wall-degrading enzymes (CWDE), such as cellulases and hemicellulases, chitinases, proteases, xylanase, pectinase, lipase, amylase, arabinase, protease, 1,3-glucanases, and lytic enzymes among other secondary metabolites (**Table 1** and **Figure 3**), which are essential for the mycoparasitism/biocontrol process [41]. These enzymes degrade the pathogen cell walls, which are composed of chitin and glucan polysaccharides. Production and regulation of lytic enzymes, which include the chitinolytic enzymes β-N-acetyl glucosaminidase, endochitinase, and chitobiosidase, are essential for promoting cell wall degradation, mycelial autolysis, chitin assimilation, and fungal parasitism and inhibiting spore germination of other plant pathogenic fungi, which are produced by *T. harzianum, T. atroviride P. karst,* and *T. asperellum* [42]. For plant protection, *Trichoderma* produces several types of volatile metabolites, such as 6-n-pentyl-2H-pyran-2-one (6-PAP) [43]. The enzymes β-1,3- and β-1,6-glucanases determine the hyperparasitic capability of *Trichoderma* in response to *Phytophthora* sp. and *Pythium* species. Endo- and exoproteases are proteolytic enzymes of *Trichoderma* that are responsible for enzyme secretion for the control of *Botrytis cinerea, Rhizoctonia solani,* and *Fusarium culmorum*. Ceratin cellulase enzymes, such as exo-β-1,4-glucanases, endo-β-1,4-glucanases, and β-glucosidases, synthesized by antagonists, also play a key role in hyperparasitism. Chemical attack

Secondary	Metabolites effect	Reference
Peptaibols	Inhibit the activity of β-1,3 glucan synthase	[36]
Gliotoxin, Gliovirin, Koninginins, Trichothecenes, and 6-Pentyl-A-Pyrone (6-PAP)	Antifungal	[37]
Trichodermin, Suzukacillin, and Alamethicin	Assist in destroying the cell walls of other pathogenic fungi	[38]
Harzianic acid, Tricholin, Massoilactone, Viridin, Glisoprenins, and Heptelidic acid	Antibiotic	[37]
Jasmonic acid and Terpenes	Repel insects from consuming plant leaves	[39]

Table 1.
Secondary metabolites produced by Trichoderma spp.

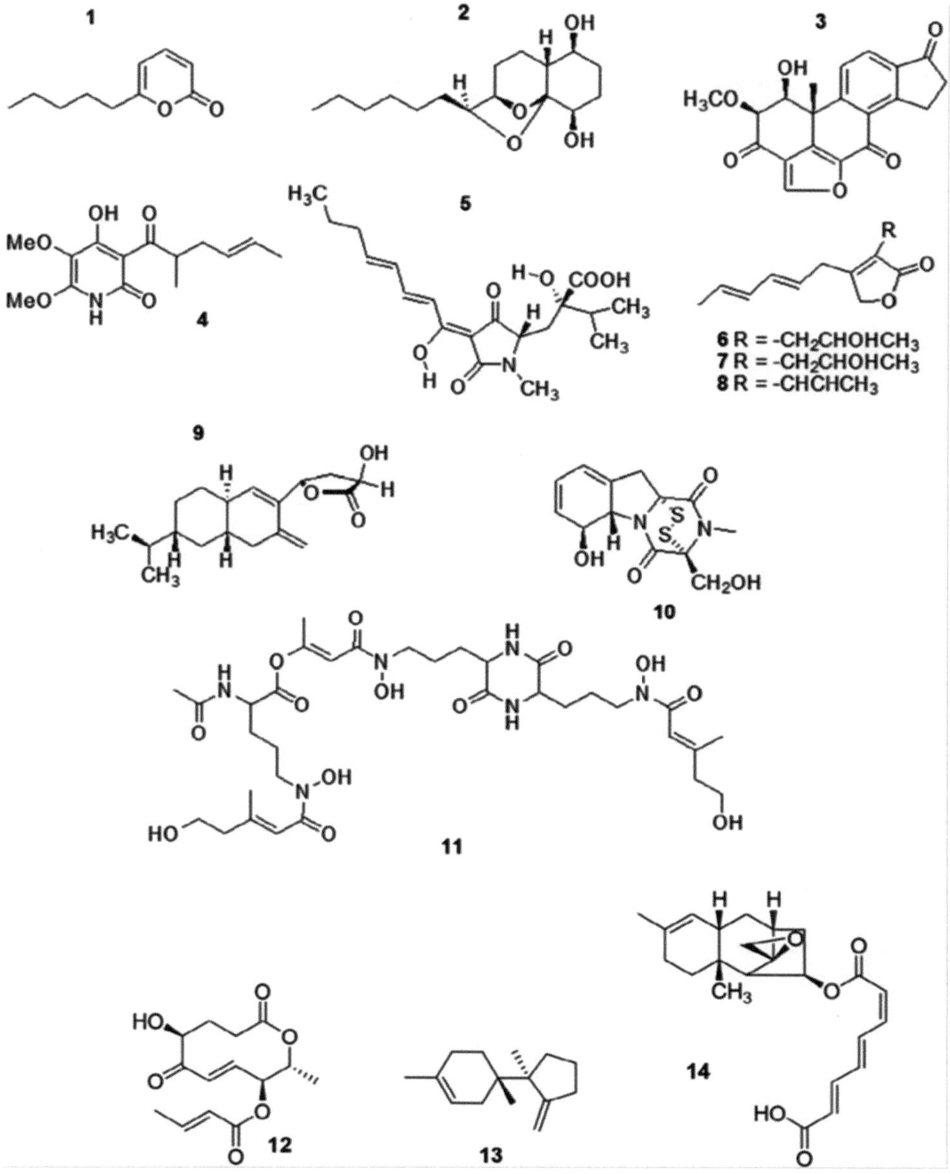

Figure 3.
Example of SMs produced by Trichoderma *spp. chemical structure of 1 6-pentyl-α-pyrone, 2 koninginin A, 3 viridin, 4 harzianopyridone, 5 harzianic acid, 6 harzianolide, 7 T39butenolide, 8 dehydro-harzianolide, 9 cerinolactone, 10 gliotoxin, 11 coprogen, 12 aspinolide C, 13 trichodiene, 14 harzianum A [40].*

and degradation of the pathogen's cell wall by hydrolytic enzymes and antifungal compounds produced by *Trichoderma* is the final stage of the mycoparasitic interaction, ultimately causing host death (**Figure 4**) [44].

Numerous *Trichoderma* genes that code for proteases and oligopeptide transporters have been identified during interactions between different *Trichoderma* species and the host. Additionally, it has been proposed that pathogenic host sensing is mediated by class IV G-protein-coupled receptors (GPCRs), which operate as sensors

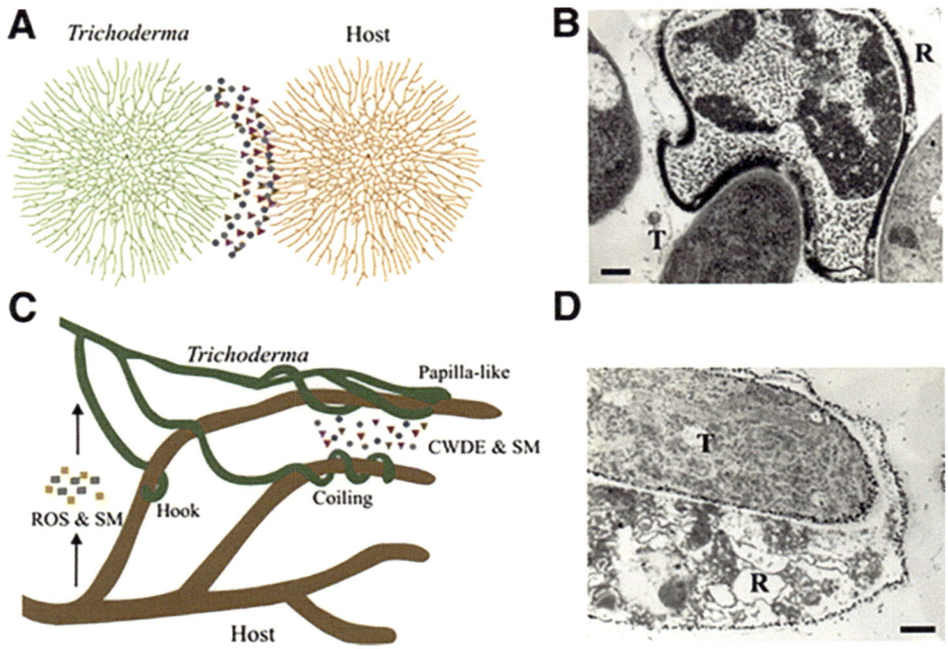

Figure 4.
Mycoparasitism by Trichoderma. (A) Interaction between colonies of Trichoderma and a fungal prey. (C) Main factors and structures used by Trichoderma during antagonistic interaction. (B and D) Host invasion. Transmission electron micrographs of T. Atroviride parasitizing Rhizoctonia solani. (B) T. Atroviride (T) penetrates R. solani (R). (D) T. atroviride (T) grows inside R. solani (R) hypha [45].

for oligopeptides and other chemicals released from phytopathogen cell walls by the activity of protease enzymes [10]. The conserved signaling cascade of G-proteins, which consists of the Gα, Gβ, and Gγ subunits, is used for further signal transduction. Mutants of *T. atroviride* that lack the Gα subunit are completely incapable of mycoparasitism, have decreased chitinolytic activity, and are unable to produce the antifungal chemical 6-pentyl-α-pyrone [10]. Furthermore, during mycoparasitism in *Trichoderma*, mitogen-activated protein kinases (MAPKs), particularly pathogenicity MAPK (TmkA and Tmk1), are implicated in signal transduction pathways [46]. The antagonistic interactions between *S. rolfsii*, *R. solani*, and *Pythium ultimatum* and *T. virens* are reduced when TmkA is removed.

Trichoderma's mycoparasitism extends beyond the host's pathogenic hyphae, and these fungi may use the pathogen's conidia as a source of further biocontrol. The *Trichoderma* TkZ3A0 strain's hyphae were able to parasitize phialides with macroconidia of the *F. culmorum* phytopathogen (isolated from winter wheat with severe fusariosis symptoms) and exhibited clear chemotaxis and adhesion to their structures. This strain has a high degree of similarity to the *T. koningiopsis* species. Even in the initial phases of the interaction, macroconidia's shape changed and their ability to germinate was inhibited (mycostasis) [47].

3.2 Competitive role of *Trichoderma*

Starvation is the most prevalent cause of mortality for all living organisms. Microorganisms typically die from starvation because there are not enough nutrients in their local surroundings, which include soil and plant surfaces. *Trichoderma* successfully established themselves in the environment, both pathogens and bio-control agents are competing hard with one another for resources like food and space. Competition is a process that occurs between microorganisms when there is a scarcity of micro- and macronutrients, and the competition for those nutrients results in the natural management of fungal communities and the development of phytopathogens [48]. Competition for micro- and macronutrients, which are carbon, nitrogen, and iron, is essential for the interactions between beneficial and harmful fungi and is associated with biocontrol systems. *Trichoderma* is usually considered as a strong competitor against soilborne fungal pathogens. It exhibits higher capacity to mobilize and quickly absorb the nutrients needed for the growth of the pathogen fungi, leading to nutritional deficit and preventing the growth and reproduction of the pathogenic fungi [49]. Rapid germination rate of *Trichoderma* has produced a wider accessible area and a greater supply of nutrients for growth. The most effective utilization of nutrients depends upon the ability of *Trichoderma* spp. to obtain energy from the metabolism of carbohydrates such as cellulose, chitin, glucan, and glucose, which are frequently present in the mycelial environment [50]. *Trichoderma* has a strong capacity for environmental adaptation [51]. It can seize nutrients and space close to the plant rhizosphere by its rapid growth and reproduction, consume oxygen in the air, and inhibit the growth of plant pathogenic fungi [52]. *Trichoderma* can effectively suppress the growth of plant pathogenic fungi since its growth rate is much faster than that of plant pathogenic fungi. After 24 hours in the soil, *Trichoderma* can quickly attach to plant roots for reproduction, and the hyphae quickly wrap plant roots to create a protective layer that prevents pathogen invasion and removes surrounding pathogens. According to Risoli et al. [53], *T. harzianum* grew 2.0–4.2 times more quickly than *B. cinerea*. *Trichoderma* mycelium competed with *Fusarium graminearum* by adhering, twining, interpenetration, and other mechanisms, causing *Fusarium graminearum*'s mycelium to become distorted and subsequently disappear [54]. Through rapid growth and reproduction, *Trichoderma* can weaken and exclude the gray mold pathogen in the same habitat by capturing water and nutrients, occupying space, using oxygen, and other things [55]. *Trichoderma* accomplishes this by biosynthesizing and releasing organic acids including gluconic, citric, and fumaric acids to cause a decrease in soil pH. Additionally, these organic acids help to dissolve micronutrients and mineral cations such phosphates, iron, manganese, and magnesium.

Siderophores, which are formed under iron-deficiency stress and have a low molecular weight (less than 10 kDa) chelator molecule with a high affinity for iron (Fe), are considered to be of the greatest significance in relation to the competitive phenomenon in *Trichoderma* [56]. The Fe ions serve as cofactors for numerous enzymes and are crucial for the healthy growth and development of both microbes and plants. Generally, the three classes of microbial siderophores—hydroxamate, catecholate, and carboxylate—are separated based on the chemical structure and the position of their coordination sites with iron [57]. The majority of the times, fungi produce hydroxamate-type siderophores, like coprogens, ferrichromes, and fusarinines, which have the structural component N5-acyl-N5-hydroxyornithine [58]. Iron is mostly found in soil in the form of Fe^{3+} under neutral pH conditions and in the presence of oxygen. Fe tends to produce insoluble iron oxides in the aerobic environment, making it unavailable to plants. Siderophores have the ability to interact with insoluble iron (Fe^{3+}) and subsequently change it into soluble Fe^{2+}, which is readily absorbed by plants and microbes (**Figure 4**).

In the absence of iron supplies from an associated niche, *Trichoderma* may inhibit the growth and activity of target soil pathogens. The interaction of the pathogens *Fusarium acuminatum, Alternaria alternata,* and *Alternaria infectoria* with *T. harzianum* was studied *in vitro*, and the results revealed that the studied fungi died from nutrient starvation. Further, the iron competition was suggested as one of the essential components in *T. asperellum's* antagonistic relationship with *F. oxysporum* f. sp. lycopersici and tomato plants' defense against wilt disease. *Trichoderma* has been shown to successfully compete with phytopathogens from the genera *Colletotrichum* sp., *Botrytis* sp., *Verticillium* sp., and *Phytophthora* sp. for simple and complex carbon substrates [59]. Further, the powerful inhibitory impact of *Trichoderma* on the pathogens *B. cinerea*, *F. graminearum*, and *Macrophomina phaseolina* was mediated by enzymes that create competition for nutrients and space [60].

Therefore, the coordination of many strategies, including the competition for nutrients, which is considered to be among the most essential, is required for the control management of some pathogens (such as *B. cinerea*) by using *Trichoderma* [61].

3.3 Induced systemic resistance of *Trichoderma*

One of the most significant mechanisms of *Trichoderma's* biocontrol actions is induced systemic resistance. It is well-known that *Trichoderma* spp. are able to induce systemic and local resistance in plants against plant pathogens through alterations in their physiological and biochemical processes, as a result of complex interactions between microbial biomolecules and plant receptors [62]. While preventing the pathogenic fungus from proliferating and growing, it can also encourage crops to produce self-defense mechanisms for obtaining local or systemic disease resistance. This causes the synthesis of regulatory plant proteins that may contribute in the defense mechanism by recognizing the microbial biomolecules, as well as the formation of the defense-enhancing chemical salicylic acid throughout the entire plant as a systematically induced defense mechanism [63]. Trichoderma-induced plant disease resistance is achieved by two methods: first, regulate the elicitors or effectors that initiate the plant disease resistance response, and second, the cell wall-degrading enzymes (such as celluloses, chitinases, and glucanases) produced by *Trichoderma* release oligosaccharides that can induce plant resistance [64]. Elicitors play a key role in the initiation of *Trichoderma*-induced systemic resistance (TISR) in plants. Currently, there are more than 10 *Trichoderma* elicitors known to promote plant resistance, including Sm1, QID74 hydrophobic protein, MRSP1, chitin-degrading enzyme, xylanase, cellulase, endopolygalacturonase, sucrase, and antibacterial peptides. These substances primarily come from five *Trichoderma* species: *T. asperellum, T. viride, T. atroviride, T. harzianum*, and *T. virens* [65–68].

The elicitors are divided into two categories: (1) race-specific elicitors that only activate gene-to-gene defense in particular host cultivars and (2) general elicitors released from pathogenic and nonpathogenic strains that activate non-race-specific defense in both host and nonhost plants. Numerous classes of elicitors have been identified, such as oligosaccharides (glucans, chitins, and oligogalacturonides), proteins and peptides (elicitins and endoxylanase), glycopeptides and glycoproteins (e.g., glycopeptide fragments of invertase), glycolipids (e.g., lipopolysaccharides), and lipophilic substances (e.g. fatty acids). In plants, the activation of signal transduction pathways by elicitors causes physical, biochemical, and molecular changes such as ion flow across the membrane, the production of reactive oxygen species (ROS), the construction of a physical barrier that inhibits the spread of phytopathogens (callose deposition and reinforcement of the plant cell wall), and the synthesis of various defense compounds (such as phytoalexins, volatile organic compounds, enzymes, and phytohormones) [69].

Numerous plant species, including monocotyledonous and dicotyledonous, have an enhanced immune response when nonpathogenic *Trichoderma* fungi are present. The recognition of conserved domains, such as the pathogen-associated molecular pattern (PAMP) or the microbe-associated molecular pattern (MAMP), is the main component of the plant defense response. Both MAMP-triggered immunity (MTI)/ PAMP-triggered immunity (PTI) and effector-triggered immunity (ETI) are induced by these domains in plants [47].

According to Saravanakumar et al. [70], *Trichoderma*-coated maize seeds significantly increased in peroxidase (POD) and phenylalanine ammonia lyase (PAL) activity, and the plants were resistant to curvularia leaf spot of maize.

3.4 Antibiosis effect of *Trichoderma*

The biological control mechanism referred to as antibiosis involves the production and excretion of secondary metabolites, including compounds of a different chemical nature with cytotoxic activity, which can reduce or inhibit the growth of phytopathogens by secreting antagonistic substances [71]. *Trichoderma* species have the ability to create antimicrobial secondary metabolites, such as trichomycin, gelatinomycin, chlorotrichomycin, and antibacterial peptides, which may inhibit the development of several types of pathogens. This inhibiting process is known as antibiosis. Antibiosis is one of the primary mechanisms by which *Trichoderma* and other biological pest controllers, such as plant growth-promoting bacteria (PGPB), work [72, 73]. More than 180 secondary metabolites of *Trichoderma* have been classified into different categories of chemical compounds based on their roles in competition, iron-chelating metabolites, inducers of plant resistance, plant growth-promoting metabolites, and antibiotics [40]. These substances can be divided into peptaibols, water-soluble substances, and volatile or nonantibiotic substances. These substances have a variety of roles in the biological control of various pathogens and are also able to function as antibacterial agents, encourage plant growth, and serve as valuable building blocks for the creation of agricultural antibiotics [74]. Some secondary metabolites have the ability to modify the metabolism and growth of plants. Some *Trichoderma* strains, including *T. viride, T. harzianum,* and *T. koningii*, have been shown to be able to produce and secrete 6-PAP, a volatile metabolite that can inhibit the growth of colonies to varying degrees, and some of them can inhibit the growth of colonies by more than 80% [75–77]. These volatile metabolites are crucial to the biocontrol of a variety of pathogenic species, including *Fusarium oxysporum, Botrytis cinerea*, and *Rhizoctonia solani* [78]. For example, *T. virens* species produce trichodermamides, whereas *T. koningii* produce Koninginins, both of which have antibacterial and antifungal activity. Additionally, in *T. harzianum* and *T. virens* substances like azaphilones, viridins, nitrogen heterocyclic compounds (such harzianopyridone and harzianic acid), and volatile terpenes have been characterized and are used in the biocontrol of pathogenic fungus [79]. Different *Trichoderma* spp. have also been studied for their ability to produce hydrolytic enzymes and proteases, including exo- and endochitinases, chitinases, xylanases, glucanases, lipases, and endo- and exopeptidases, among others, with antifungal activity [80]. The most prevalent volatile organic compound (VOC) from *T. atroviride* is 6-pentyl-2H-pyran-2-one (6-PP), which, together with other VOCs produced by the fungus, promotes plant growth and controls sugar transport in *Arabidopsis* roots [81].

4. *Trichoderma* as biocontrol agent against soilborne pathogen

The rapid use of chemical fertilizers considerably contributes to the current condition of environmental deterioration through the use of fossil fuels, the production and release of carbon dioxide, and the contamination of water supplies. Globally, environmental deterioration is currently a major concern. The most effective way to prevent environmental deterioration is by using biological agents [82]. Biocontrol is the use of biological populations to control the proliferation of pests. A few of the biocontrol mechanisms by which BCAs can prevent the growth of soilborne pathogens include the capability to grow faster than soilborne pathogens for nutrients and space, the production of numerous potent plant-degrading enzymes like lytic enzymes and proteolytic enzymes, and the production of more than 200 antibiotics that are extremely toxic to all macro- and microorganisms. It is believed that the ability to produce various antibiotics will enhance biological control by preventing a variety of microbial competitors, some of which are likely plant diseases [28].

Antibiotics such as 2,4-diacetylphloroglucinol, produced by *Pseudomonas fluorescens* F113 against *Pythium* species, which causes damping-off disease, and gliotoxin, produced by *Trichoderma virens* against *Rhizoctonia solani*, which causes plant root rot, have been reported to be involved in the suppression of plant pathogens. Furthermore, BCA enzymes can also hydrolyze chitin, proteins, cellulose, and hemicellulose. As a result, plant pathogens are quickly inhibited. Several types of BCAs can produce enzymes that are beneficial against particular plant diseases [83].

Trichoderma has a significant application value and potential for biological management of plant diseases. The use of *Trichoderma* to control plant diseases has been studied all around the world. *Trichoderma* spp. are effective biocontrol agents that are widely used against soilborne diseases. Some of the important gene types that are essential to the function of biocontrol include protease, chitinase, glucanase, tubulins, cell adhesion proteins, and stress-tolerant genes. These genes play a role in the degradation of cell walls, hyphal growth, stress tolerance, and parasitic activity. For example, chitinase breaks down glycosidic linkages, while xylanase breaks down hemicellulose and so forth. *Trichoderma* fights against other plant pathogenic fungi and promotes the growth of the plant and its roots. For plant pathogenic diseases, it uses a variety of management strategies, such as antibiosis, mycoparasitism, induced host cell resistance, and competition for nutrients and space. The antagonist's proliferative potential in the soil is a key factor in the efficient biocontrol of soilborne pathogens. *Trichoderma* species are recognized as prospective biological control agents due to their capacity to either increase resistance or restrict the growth of a number of phytopathogenic fungus. *Trichoderma's* induction of systemic resistance is another important method of pathogen resistance. Through mycoparasitism, competition, or the production of antibiotics, plants can defend themselves directly [84].

As a biocontrol agent, *T. viride* and *T. harzianum* are known to have various degrees of inhibitory effects on pathogens including *Fusarium oxysporum, Rhizoctonia solani, Pythium aphanidermatum, Fusarium culmorum, Gaeumannomyces graminis var. tritici, Sclerotium rolfsii, Phytophthora cactorum, Botrytis cinerea, Pseudocercospora* spp., *Colletotrichum* spp., and *Alternaria* species [85–90]. *Trichomonas* has been widely used in the biological management of tobacco root rot, tomato, potato, rice, wheat, and other plant diseases (**Table 2**).

Disease	Crop	Pathogen	Biocontrol strain	References
Root rot disease	Soybean (*Glycine max* (L.) Merr. cv)	*Pythium arrhenomanes f. sp. adzuki*	*T. viride*	[91]
	Corn (*Zea mays*)	*Fusarium oxysporum f. sp. adzuki*		
	Cocoyam (*Xanthosoma sagittifolium*)	*Pythium myriotylum*	*T. asperellum*	[92]
	Pepper plants (*Capsicum annuum*)	*Rhizoctonia solani*	*T. harzianum*	[93]
	Eggplant (*Solanum melongena* L.)	*Macrophomina phaseolina*	*T. harzianum, T. polysporum, T. viride*	[94]
Damping-off	Pepper (*Capsicum annuum*)	*Phytophthora capsici*	*T. harzianum*	[95]
	Cucumber (*Cucumis sativus*)	*Pythium sp.*	*T. harzianum*	[96]
	Cotton (*Gossyphtm hirsutum*)	*Rhizoctonia solani*	*T. harzianum*	[97]
	Cotton (*Gossyphtm hirsutum*)	*Pythium aphanidermatum, Pythium ultimum, Rhizopus oryzae*	*T. virens*	[98]
Wilt	Tomato (*Solanum lycopersicum*)	*Fusarium oxysporum f. sp. Lycopersici (FOL)*	*T. asperellum*	[99]
	Melon (*Cucumis melo*)	*F. oxysporum*	*T. harzianum T-78*	[100]
Fruit rot	Chili (*Capsicum annuum*)	*Alternaria tenuis*	*T. harzianum*	[101]
	Tomato (*Solanum lycopersicum*)	*Rhizoctonia solani*	*T. viride, T. virens, T. harzianum*	[102]
Brown spot	Tobacco (*Nicotiana tabacum*)	*Alternaria alternata*	*T. harzianum*	[103]
Brown root rot	Peanut (*Arachis hypogaea*)	*Fusarium solani*	*T. harzianum*	[104]
Anthracnose gray mold	Strawberry (*Fragaria ananassa*)	*Colletotrichum acutatum, Botrytis cinerea*	*T. hamatum, T. atroviride, T. longibrachiatum*	[105]
Head blight	Wheat and other small grain cereals (*Triticum aestivum*)	*Fusarium graminearum, Fusarium culmorum*	*T. gamsii*	[106]
Sheath blight	Rice (*Oryza sativa*)	*Rhizoctonia solani*	*T. harzianum*	[107]
Blossom blight	Alfalfa (*Medicago sativa*)	*Sclerotinia sclerotiorum*	*T. atroviride*	[108]
Web blight	Bean (*Phaseolus vulgaris*)	*Sclerotinia sclerotiorum*	*T. viride*	[102]
Collar rot	Tomato (*Solanum lycopersicum*)	*Sclerotium rolfsii*	*T. virens, T. harzianum*	[102]

Table 2.
Various diseases controlled by Trichoderma spp.

5. *Trichoderma* as biofertilizers

Severe environmental conditions are one of the factors that have decreased crop growth and productivity. Crop productivity would suffer as a result of abiotic stresses caused by climate changes, such as temperature rise, increased carbon dioxide levels, protracted drought, and hurricanes [109]. Additionally, these circumstances have an impact on the proliferation and distribution of plant pathogens and pests, adding to the biotic stresses imposed on crops [110]. Genetically uniform improved crop are frequently vulnerable to invasion caused by non-native pests and pathogens. Irrigation practices in dry environments affect soil nutrients and increase salinization. Salinization has been associated to decreased soil microbial activity and also has an impact on physical properties of soil, such as producing compaction. Compacted soil with low oxygen content affects root growth, which limits nutrient and water intake and reduces the production of plants. As a result, reports indicate that the average production of important crops is facing losses of up to 50%. In order to maintain agricultural production and productivity under various environmental challenges, *Trichoderma* spp. application as biofertilizer offers a sustainable and environmentally friendly alternative. In addition, they also play an essential role as a biocontrol agent, protecting plants from a variety of disease attacks. *Trichoderma* spp. interacts with plants to promote root growth, assist in nutrient uptake, and enhance the plant's resistance to abiotic stresses, all of which increase the production of agricultural products [47]. According to previous studies, *Trichoderma* was able to colonize roots and affect the transcriptome, proteome, and epigenome of plants, improving their capacity to tolerate external stresses. In the presence and absence of abiotic stress factors, *Trichoderma* spp. treatment has positive effects on plants (**Table 3**).

5.1 Plant growth enhancement by *Trichoderma* spp.

Trichoderma frequently coexists with the root ecosystems of the host plants. Therefore, *Trichoderma* is generally referred to as a genus of symbiotic, opportunistic, and nonpathogenic microorganisms that colonize the roots and regulate plant pathogens. *Trichoderma* spp. not only inhibit the growth of pathogens but also increase plant growth, serve as a biofertilizer, and stimulate plant defense mechanisms. According to Hyakumachi and Kubota [112], plant growth-promoting fungi (PGPF) are fungi that can play a role in the growth of plants. The main effects of PGPF are frequently seen in the production, quality, and growth of the crop. Recent studies have shown that *Trichoderma* spp. can make a great PGPF. It has been shown that some *Trichoderma* strains may penetrate the epidermis and create a robust and persistent colonization of root surfaces. *Trichoderma* species have positive impacts on plants, such as promoting growth, enhancing seed germination and viability, improving root structure and condition, and increasing photosynthetic efficiency, blooming, and yield quality [113]. *Trichoderma* has been shown to enhance the growth of lettuce, pepper plants, and tomatoes. The production of phytohormones and phytoregulators is the most significant stimulating factor at almost all stages of plant growth and development by creating a favorable environment and production of a large amount of secondary metabolites [47]. *Trichoderma* spp. produce secondary metabolites that regulate plant growth, such as koningin A (*Trichoderma koningii*) and 6-pentyl-alpha-pyrone (*T. harzianum*). Other significant activities carried out by *Trichoderma* spp. include increased solubilization of phosphate minerals (such as Fe, Mg, and Mn), reduction in soil pH, and formation of gluconic and citric acids as well as micronutrients.

The main advantage of *Trichoderma* spp. for plant growth is the development of roots. This theory is confirmed by recent studies [114] that have shown *Trichoderma* spp. produces or controls plant hormones like auxin, harzianic acid,

Strain as biofertilizer	Crops	Application mode	Beneficial outcome
T. azevedoi	Lettuce	Simple exposure	Increases carotenoids and chlorophyll with reduction in the white mold attack to about 78.83%
T. afroharzianum	Tomato	Seed inoculation or treatment	Helps in the secretion of phytohormones like homeostasis, antioxidant activity, phenylpropanoid biosynthesis, and glutathione metabolism
T. harzianum, *T. asperellum*, *T. hamatum*, *T. atroviride*	Chinese cabbage	Irrigation	Increased yield by 37%; Increased enzyme activity in the soils and providing more inorganic nitrogen and phosphorus content to the soil
T. brevicompactum, *T. gamsii*, *T. harzianum*,	Tomato	Seedling drenching	Improved growth and yield due to the production of indole-3 acetic acid
T. harzianum Rifai,	Tomato	Seed treatment	Improves phosphorus uptake
T. asperellum,			
T. brevicompactum, *T. gamsii*, *T. harzianum*	Tomato	Seed drenching	Improves phosphorus solubilization
T. erinaceum	Rice	Seed treatment	Improves germination, vigor, and yield
T. harzianum T22	Tomato	Seed treatment	Improves soil fertility, level of minerals and antioxidants, nutrient uptake, and yield
T. harzianum T22	Tomato	Soil amendment as compost	Increase in yield to about 12.9%
T. viride	Sugar cane	Powder as fertilizers	Improves nutrient uptake
T. asperellum T34	Cucumber	Seedling drenching	Enhanced nutrient uptake
T. harzianum	All crops	Compost	Enhances residue decomposition resulting in availability of soil nutrients
T. simmonsii	Bell pepper	Seedling drenching	Improves yield to about 67%
T. reesei	Chickpea	Seed treatment	Enhanced mineral uptake
T. harzianum	Mustard	Soil inoculation	Improved nitrogen absorption and increased yield to about 108 and 203%
T. harzianum	Chili	Soil inoculation	Increased yield
T. harzianum	Barley	Seed inoculation	17% increase in yield
T. viridae	Wheat	Soil and seed inoculation	75.8% increase in yield with improved nutrient absorption
T. viridae	Potato	Soil inoculation	Increased yield with an average of 16.25 tubers/plant
T. viridae	Red beet cabbage	Seed inoculation	29% increase in yield

Strain as biofertilizer	Crops	Application mode	Beneficial outcome
T. harzianum	Onion seedlings	Seedling inoculation	Enhanced growth and yield
T. asperellum	Maize	Soil granules	Improves yield

Table 3.
Trichoderma spp. as biofertilizers function in increasing plant production and growth [111].

and harzionalide, which are responsible for promoting root development. According to Yedidia et al. [115], it was found that plants inoculated with *T. harzianum* on the 28th day developed significantly greater roots. Additionally, this study found that the content of Cu, P, Fe, Zn, Mn, and Na increased in the inoculated root (**Figure 5**). At the same time, it was found that the concentration of Mn, Zn, and P had increased by 70%, 25%, and 30%, respectively, in the plant's shoot. The study supports the theory made by Contreras-Cornejo et al. [116] that a high root surface area caused by *Trichoderma* spp. inoculation permits the root to explore a larger area of soil. This makes it possible for plants to absorb more macronutrients and micronutrients from the soil, which is advantageous for them when fighting for minerals with other organisms or when there is a lack of minerals. The identification of *Trichoderma* spp. metabolites could be helpful in the application of novel biofertilizers in agriculture as an alternative to synthetic/chemical fertilizers. To increase crop yield, using *Trichoderma* spp. as a biofertilizer in combination with fungi as inoculants may be beneficial. Additionally, it reduces pollution caused by the overuse of synthetic/chemical fertilizer in the agricultural sector (**Figure 6**).

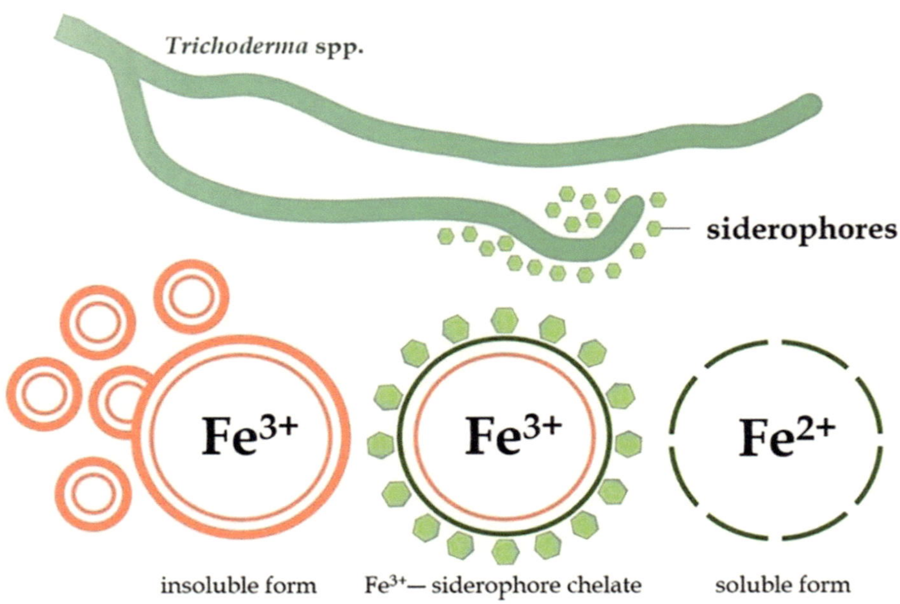

Figure 5.
Transformation of the insoluble form of iron (Fe^{3+}) into a soluble and easily assimilating form (Fe^{2+}) by siderophores produced by Trichoderma fungi [117].

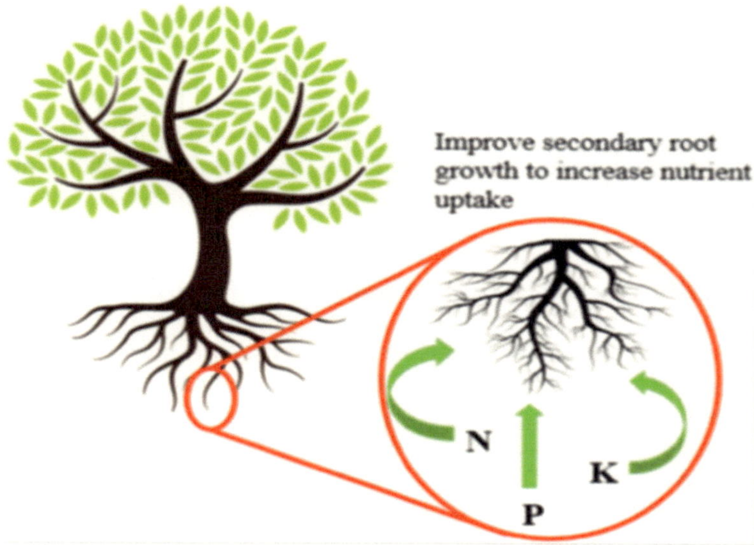

Figure 6.
Trichoderma spp. increase the nutrients uptake by root improvement [118].

5.2 Plant root colonization by *Trichoderma* Spp.

Many rhizosphere *Trichoderma* species may colonize the root surfaces of monocotyledonous and dicotyledonous plants, which may have a significant effect on the plant's metabolism. *Trichoderma* species recognize the host plant and adhere and penetrate plant roots to colonize it. *Trichoderma* spp. colonization of plant roots enhances plant defense by causing the production of 1,3-glucanase, peroxidases, phenylalanine, chitinases, and hydroperoxidase, which activated signaling of the plant's biosynthetic pathways and led to the accumulation of low-molecular weight phytoalexins. Since it colonized such a diverse range of host plants, Trichoderma most likely developed efficient techniques to get over plant defenses. *Trichoderma harzianum* and plant interacted physically in a symbiotic way where *Trichoderma* stimulate the increased activity of metabolites such as peroxidase and chitinase, thus protecting the plant from disease in exchange for it providing a nutritional niche for the fungus. Yedidia et al. [119] used an electron microscope to investigate the physical interaction between *T. harzianum* T-203 and a cucumber plant and discovered that the fungus pierced the root and spread across the epidermis and outer cortex, which prompted the production of more peroxidase and chitinase. Consequently, the interaction seems to be a symbiotic one in which *Trichoderma* lived in the nutritional niche offered by the plant and the plant was protected from disease.

Various studies have confirmed *Trichoderma's* capacity to colonize the root surface in addition to the rhizosphere soil. Previously, the root hair and elongation zone was the main area where *Trichoderma* colonized the roots. However, recent research has shown that these fungi have colonized the root cap border cells (RBCs) zone. RBCs play a vital role in the interaction between plants and soil microbes. During root extension, the RBCs are discharged into the soil environment from the cap cells' outer layer. These cells are thought to be a valuable source of nutrients and biologically active substances for microorganisms, along with the related root exudates. Additionally,

RBCs have an impact on the structure and content of the microbial rhizosphere population and serve as attractants for microbes, promoting interaction with plants and root colonization. RBCs are also considered to be an important factor in protecting plants from a variety of biotic and abiotic stresses [120]. The ability of *Trichoderma* to colonize RBCs, as shown for the wheat roots injected with DEMTkZ3A0 strain conidia, suggests a major role of these fungi in the indirect defense mechanism against phytopathogens through the creation of a mantle-like structure.

6. Commercially available *Trichoderma*-based bioproducts

Trichoderma spp. are considered as one of the most extensively studied fungal agents as microbial biocontrol agents (MBCAs) in agriculture and are sold as biopesticides, biofertilizers, growth promoters, and naturally occurring resistance-inducing agents. Since these compounds are developed to protect plants from adverse environmental conditions, it is important to establish set standards for their production and maintenance because changes in pH, temperature, and other environmental factors may have an impact on the formulation's activity. Additionally, it should not be phytotoxic, be economical, dissolve quickly in water, be easy and affordable to obtain carrier materials for, and be compatible with agrochemicals. The distribution of the different *Trichoderma* spp.-based products for important agricultural diseases is listed in **Table 4**.

Bioproducts name	Species strain	Company, Country
Topshield, Rootshield	*T. harzianum T-22*	Bioworks, Geneva, N.Y.
T35	*T. harzianum*	Makhteshim-Agan Chemicals, Israel
Harzian 20, Harzian 10	*T. harzianum*	Natural Plant Protection, Noguerres, France
F-stop	*T. harzianum*	Eastman Kodak Co., United States TGT Inc., New York
Supraavit	*T. harzianum*	Bonegaard and Reitzel, Denmark
Solsain, Hors-solsain, Plantsain	*Trichoderma spp.*	Prestabiol, Montpellier, France
ANTI-FUNGUS	*Trichoderma spp.*	Grondontsmettingen De Ceuster, Belgium
Ty	*Trichoderma spp.*	Mycontrol, Israel
GlioGard and SoilGard	*T. virens (Gliocladium virens)*	Grace-Sierra Co., Maryland, USA
Bip T	*T. viride*	Poland
Promot Plus WP Promot PlusDD	*Trichoderma spp., T. koningii, T. harzianum*	Tan Quy, Vietnam
TRiB1	*Trichoderma spp.*	National Institute of Plant, Vietnam
TRICÔ-DHCT	*Trichoderma spp.*	Can Tho University, Vietnam
Vi – DK	*Trichoderma spp.*	Pesticide Corp., Vietnam
NLU-Tri	*Trichoderma virens*	Ho Chi Minh University of Agriculture and Forestry, Vietnam
Biobus 1.00WP	*Trichoderma viride*	Nam Bac, Vietnam

Bioproducts name	Species strain	Company, Country
Bio – Humaxin Sen Vàng 6SC, Fulhumaxin 5.15SC	*Trichoderma spp.*	An Hung Tuong, Vietnam
BioSpark Trichoderma	*T. parceramosum, T. pseudokoningii, and Ultraviolet irradiated strain of T. harzianum*	BioSpark Corporation, Philippines
Biocure F	*T. viride*	T. Stanes and Company Limited, European Union, available in India
Trichoderma viride powder	*T. viride*	Organic Dews, Bio Organic, India
Bio-Shield, Bioveer	*T. viride*	Ambika Biotech, India
Mycofungicyd, Trichodermin	*T. viride 16, T. lignorum*	Bizar-agro LTD, Ukraine
ICB Nutrisolo SC e WP	*T. viride, T. harzianum, T. koningii* and *Trichoderma* spp.	ICB BIOAGRITEC Ltd., Brazil
Trichosav-34	*T. harzianum* A-34	Institute for Research in Plant Protection (INISAV), Cuba
Trichosav-55	*T. harzianum* A-55	Institute for Research in Plant Protection (INISAV), Cuba
Antagon TV	*T. viride*	Green Tech Agroproducts, Tamil Nadu, India
Trichostar	*T. harzianum*	Green Tech Agroproducts, Tamil Nadu, India
Gliostar	*T. virens*	GBPUAT, Pantnagar, India
Bioderma	*T. viride, T. harzianum*	Biotech International Ltd., India
Bio Fit	*T. viride*	Ajay Biotech (India) Ltd., India
Ecofit	*T. viride*	Hoechst Schering Afgro Evo Ltd., India
Trichoguard	*T. viride*	Anu Biotech Int. Ltd. Faridabad, India
Biocon	*T. viride*	Tocklai Experimental Station Tea Research Association, Jorhat (Assam), India

Table 4.
Description of commercially available bioactive products of Trichoderma spp. worldwide [111].

7. Conclusion

Currently, the primary strategy for controlling plant diseases is chemical control, which is accomplished by spraying pesticides and fungicides. Although chemical control has a positive effect and is beneficial in boosting agricultural production, it also has negative effects on people and the environment and increases pathogens' resistance. Scientists and their studies have demonstrated that *Trichoderma* is non-pathogenic to plants and exhibits various biocontrol features, which makes it one of the most efficient organisms against several types of plant pathogens, and reduces the use of harmful pesticides in the agricultural sector. *Trichoderma* spp., which are primarily found in soil and the rhizosphere region, are antagonistic to the majority of plant pathogens. A biocontrol program is only established when the biocontrol agent is able to successfully control the relationship between the host plant and pathogen. *Trichoderma* has a well-established ability to control this interaction. It has also been shown that the fungi improve plant defense mechanisms. Furthermore, *Trichoderma* provides a variety of control strategies for various plant pathogens, which gives it an edge over other phytopathogen control strategies. Mechanisms that are usually involved are mycoparasitism, antibiotics, competition for nutrients, and stimulation of systemic resistance in plants. Another fascinating feature is that *Trichoderma* species produce a number of secondary metabolites that significantly affect plant growth or prevent pathogen growth along with promoting localized and systemic resistance as well as stress tolerance in plants. Currently, *Trichoderma* species are being used in the sustainable disease management system for managing plant diseases. These methods of managing plant diseases may be safer for the agroecosystem, overall environment, the health of the soil, and human health. They would also be the proper move in sustaining agricultural output to satisfy the demands of growing populations. As a result, farmers have the choice of using it to enhance agricultural yields and produce higher quality. *Trichoderma* spp. can be used for waste/organic material decomposition, contaminated area purification, and decreasing diseases and enhancing plant growth.

Author details

Lovely Bharti*, Kajol Yadav and Ashok Kumar Chaubey
Nematology Laboratory, Department of Zoology, Chaudhary Charan Singh University Meerut, UP, India

*Address all correspondence to: lovely13589@gmail.com

References

[1] Alizadeh M, Vasebi Y, Safaie N. Microbial antagonists against plant pathogens in Iran: A review. Open Agriculture. 2020;**5**:404-440

[2] Ghorbanpour M, Omidvari M, Abbaszadeh-Dahaji P, Omidvar R, Kariman K. Mechanisms underlying the protective effects of beneficial fungi against plant diseases. Biological Control. 2018;**117**:147-157

[3] Grasswitz TR. Integrated pest management (IPM) for small-scale farms in developed economies: Challenges and opportunities. Insects. 2019;**10**:179

[4] Gilardi G, Manker DC, Garibaddi A, Gullino ML. Efficacy of the biocontrol agents *Bacillus subtilis* and *Ampebmyces quisqualis* applied in combination with fungicides against powdery mildew of *zucchini*. Journal of Plant Disease and Protection. 2008;**115**:208-213

[5] Naher L, Yusuf UK, Siddiquee S, Ferdous J, Rahman MA. Effect of media on growth and antagonistic activity of selected *Trichoderma* strains against *Ganoderma*. African Journal of Microbiology Research. 2012;**6**:7449-7453

[6] Contreras-Cornejo HA, Macías-Rodríguez L, del-Val E, Larsen J. Ecological functions of *Trichoderma* spp. and their secondary metabolites in the rhizosphere: Interactions with plants. FEMS Microbiology Ecology. 2016;**92**(4):fiw036. DOI: 10.1093/femsec/fiw036

[7] Hermosa R, Viterbo A, Chet I, Monte E. Plant beneficial effects of *Trichoderma* and of its genes. Microbiology. 2012;**158**:17-25

[8] Shoresh M, Harman GE, Mastouri F. Induced systemic resistance and plant responses to fungal biocontrol agents. Annual Review of Phytopathology. 2010;**48**:21-43

[9] Harman GE, Howell CR, Viterbo A, Chet I, Lorito M. *Trichoderma* plant symbionts species-opportunistic, avirulent plant symbionts. Nature Reviews. Microbiology. 2004;**2**:43-56

[10] Druzhinina IS, Seidl-Seiboth V, Herrera-Estrella A, Horwitz BA, Kenerley CM, Monte E, et al. *Trichoderma*: The genomics of opportunistic success. Nature Reviews. Microbiology. 2011;**9**:749-759. DOI: 10.1038/nrmicro2637

[11] Sharon E, Bar-Eyal M, Chet I, Herrera-Estrella A, Kleifeld O, Spiegel Y. Biological control of root-knot nematode *Meloidogyne javanica* by *Trichoderma harzianum*. Phytopathology. 2001;**91**(7):687-693

[12] Yasmeen R, Siddiqui ZS. Physiological responses of crop plants against *Trichoderma harzianum* in saline environment. Acta Botanica Croatica. 2017;**76**(2):154-162

[13] Brotman Y, Landau U, Cuadros-Inostroza L, Takayuki T, Fernie AR, Chet I, et al. *Trichoderma*-plant root colonization: Escaping early plant defense responses and activation of the antioxidant machinery for saline stress tolerance. PLoS Pathogens. 2013;**9**(3):e1003221. DOI: 10.1371/journal.ppat.1003221

[14] Kamala T, Devi SI, Sharma KC, Kennedy K. Phylogeny and taxonomical investigation of *Trichoderma* spp. from Indian region of Indo-Burma biodiversity hot spot region with special reference to Manipur. BioMed Research

International. 2015;**2015**:285261. DOI: 10.1155/2015/285261

[15] Błaszczyk L, Siwulski M, Sobieralski K, Lisiecka J, J˛edryczka M. *Trichoderma* spp.—Application and prospects for use in organic farming and industry. Journal of Plant Protection Research. 2014;**54**(4):309-317

[16] Sánchez-Montesinos B, Santos M, Moreno-Gavíra A, Marín-Rodulfo T, Gea FJ, Diánez F. Biological control of fungal diseases by *Trichoderma aggressivum* f. europaeum and its compatibility with fungicides. Journal of Fungi. 2021;**7**:598. DOI: 10.3390/jof7080598

[17] Sun J, Karuppiah V, Li Y, Pandian S, Kumaran S, Chen J. Role of cytochrome P450 genes of *Trichoderma atroviride* T23 on the resistance and degradation of dichlorvos. Chemosphere. 2022;**290**:133173. DOI: 10.1016/j.chemosphere.2021.133173

[18] De Padua JC, dela Cruz TEE. Isolation and characterization of nickel-tolerant Trichoderma strains from marine and terrestrial environments. Journal of Fungi. 2021;**7**:591. DOI: 10.3390/jof7080591

[19] Escudero-Leyva E, Alfaro-Vargas P, Muñoz-Arrieta R, Charpentier-Alfaro C, Granados-Montero MM, Valverde-Madrigal KS, et al. Tolerance and biological removal of fungicides by *Trichoderma* species isolated from the endosphere of wild Rubiaceae plants. Frontiers in Agronomy. 2022;**3**:772170. DOI: 10.3389/fagro.2021.772170

[20] Weindling R. *Trichoderma lignorum* as a parasite of other soil fungi. Phytopathology. 1932;**22**:837-845

[21] Kumar MA, Sharma P. Morphological characterization of biocontrol isolates of *Trichoderma* to study the correlation between morphological characters and biocontrol efficacy. International Letters of Natural Science. 2016;**55**:57-67

[22] Di Marco S, Metruccio EG, Moretti S, Nocentini M, Carella G, Pacetti A, et al. Activity of *Trichoderma asperellum* strain ICC 012 and *Trichoderma gamsii* strain ICC 080 toward diseases of esca complex and associated pathogens. Frontiers in Microbiology. 2022;**12**:813410. DOI: 10.3389/fmicb.2021.813410

[23] Druzhinina IS, Chenthamara K, Zhang J, Atanasova L, Yang D, Miao Y, et al. Massive lateral transfer of genes encoding plant cell wall-degrading enzymes to the mycoparasitic fungus *Trichoderma* from its plant- associated hosts. PLoS Genetics. 2018;**14**:e1007322. DOI: 10.1371/journal.pgen.1007322

[24] Kubicek CP, Steindorff AS, ChenthamaraK,ManganielloG,HenrissatB, Zhang J, et al. Evolution and comparative genomics of the most common *Trichoderma* species. BMC Genomics. 2019;**20**:485. DOI: 10.1186/s12864-019-5680-7

[25] Rush TA, Shrestha HK, Gopalakrishnan Meena M, Spangler MK, Ellis JC, Labbé JL, et al. Bioprospecting *Trichoderma*: A systematic roadmap to screen genomes and natural products for biocontrol applications. Frontiers in Fungal Biology. 2021;**2**:716511. DOI: 10.3389/ffunb.2021.716511

[26] Singh HB, Singh BN, Singh SP, Singh SR, Sarma BK. Biological control of plant diseases: Status and prospects. In: Recent Advances in Biopesticides: Biotechnological Applications. Vol. 322. New Delhi, India: New India Pub.; 2009

[27] Jaklitsch WM. European species of Hypocrea part I: The green-spored species. Studies in Mycology.

2009;**63**:1-91. DOI: 10.3114/sim.2009.63.01

[28] Zin NA, Badaluddin NA. Biological functions of *Trichoderma* spp. for agriculture applications. Annals of Agricultural Science. 2020;**65**:168-178. DOI: 10.1016/j.aoas.2020.09.003

[29] Santoro PH, Cavaguchi SA, Alexandre TM, Zorzetti J, Neves PMOJ. In vitro sensitivity of antagonistic *Trichoderma atroviride* to herbicides. Brazilian Archives of Biology and Technology. 2014;**57**:238-243

[30] Kim CS, Park MS, Kim SC, Maekawa N, Yu SH. Identification of *Trichoderma*, a competitor of shiitake mushroom (Lentinula edodes), and competition between Lentinula edodes and *Trichoderma* species in Korea. Plant Pathology Journal. 2012;**28**:137-148

[31] Hatvani L, Kredics L, Allaga H, Manczinger L, Vágvölgyi C, Kuti K, et al. First report of *Trichoderma aggressivum* f. aggressivum green mold on Agaricus bisporus in Europe. Plant Disease. 2017;**101**:1052. DOI: 10.1094/PDIS-12-16-1783-PDN

[32] Chaudhari PJ, Shrivastava P, Khadse AC. Substrate evaluation for mass cultivation of *Trichoderma viride*. Asiatic Journal of Biotechnology Resources. 2011;**2**:441-446

[33] Shah S, Nasreen S, Sheikh PA. Cultural and morphological characterization of *Trichoderma* spp. associated with green mold disease of *Pleurotus* spp. in Kashmir. Research Journal of Microbiology. 2012;7:139-144

[34] Moreno-Ruiz D, Lichius A, Turra D, Di Pietro A, Zeilinger S. Chemotropism assays for plant symbiosis and mycoparasitism related compound screening in *Trichoderma atroviride*. Frontiers in Microbiology. 2020;**11**:601251. DOI: 10.3389/fmicb.2020.601251

[35] Viterbo A, Horwitz BA. Mycoparasitism. In: Borkovich KA, Ebbole DJ, editors. Cellular and Molecular Biology of Filamentous Fungi. Vol. 42. Washington: American Society for Microbiology; 2010. pp. 676-693

[36] Nawrocka J, Małolepsza U. Diversity in plant systemic resistance induced by *Trichoderma*. Biological Control. 2013;**67**(2):149-156. DOI: 10.1016/j.biocontrol.2013.07.005

[37] Kolli SK, Adusumilli N. *Trichoderma*—Its paramount role in agriculture. In: Singh J, Gehlot P, editors. New and Future Developments in Microbial Biotechnology and Bioengineering. Amsterdam, The Netherlands: Elsevier; 2020. pp. 69-83. DOI: 10.1016/B978-0-12-821007-9.00007-3

[38] Manibhushanrao K, Sreenivasaprasad S, Baby U, Joe Y. Susceptibility of rice sheath blight pathogen to mycoparasites. Current Science. 1989;**58**(9):515-518

[39] Contreras-Cornejo HA, Macías-Rodríguez L, del-Val E, Larsen J. The root endophytic fungus *Trichoderma atroviride* induces foliar herbivory resistance in maize plants. Applied Soil Ecology. 2018;**124**:45-53. DOI: 10.1016/j.apsoil.2017.10.004

[40] Vinale F, Sivasithamparam K. Beneficial effects of *Trichoderma* secondary metabolites on crops. Phytotherapy Research. 2020;**34**:2835-2842

[41] Strakowska J, Błaszczyk L, Chełkowski J. The significance of cellulolytic enzymes produced by

Trichoderma in opportunistic lifestyle of this fungus. Journal of Basic Microbiology. 2014;**54**(Suppl.1):S2-S13. DOI: 10.1002/jobm.201300821

[42] Saravanakumar K, Li Y, Yu C, Wang Q, Wang M, Sun J, et al. Effect of *Trichoderma harzianum* on maize rhizosphere microbiome and biocontrol of Fusarium stalk rot. Scientific Reports. 2017;**7**:1771

[43] Jeleń H, Błaszczyk L, Chełkowski J, Rogowicz K, Strakowska J. Formation of 6-n-pentyl-2H-pyran-2-one (6-PAP) and other volatiles by different *Trichoderma* species. Mycological Progress. 2013;**13**(3):589-600. DOI: 10.1007/s11557-013-0942-2

[44] Seidl-Seiboth V, Ihrmark K, Druzhinina IS, Karlsson M. Molecular evolution of *Trichoderma chitinases*. In: Gupta VK, Schmoll M, Herrera-Estrella A, Upadhyay RS, Druzhinina I, Tuohy MG, editors. Biotechnology and Biology of *Trichoderma*. Oxford, UK: Elsevier; 2014. pp. 67-78

[45] Guzmán-Guzmán P, Porras-Troncoso MD, Olmedo-Monfil V, Herrera-Estrella A. *Trichoderma* species: Versatile plant symbionts. Review. Phytopathology. 2019;**109**:6-16. DOI: doi.org/10.1094/phyto-07-18-0218-rvw

[46] Schmoll M. The information highways of a biotechnological workhorse—Signal transduction in *Hypocrea jecorina*. BMC Genomics. 2008;**9**:430

[47] Jaroszuk-Ściseł J, Tyśkiewicz R, Nowak A, Ozimek E, Majewska M, Hanaka A, et al. Phytohormones (auxin, gibberellin) and ACC deaminase in vitro synthesized by the mycoparasitic *Trichoderma* DEMTkZ3A0 strain and changes in the level of auxin and plant resistance markers in wheat seedlings inoculated with this strain conidia. International Journal of Molecular Sciences. 2019;**20**:4923

[48] Poveda J, Baptista P. Filamentous fungi as biocontrol agents in olive (*Olea europaea* L.) diseases: Mycorrhizal and endophytic fungi. Crop Protection. 2021;**146**:105672. DOI: 10.1016/j.cropro.2021.105672

[49] Halifu S, Deng X, Song X, Song R, Liang X. Inhibitory mechanism of *Trichoderma virens* ZT05 on *Rhizoctonia solani*. Planning Theory. 2020;**9**:912. DOI: 10.3390/plants9070912

[50] Mahmood A, Kataoka R. Potential of biopriming in enhancing crop productivity and stress tolerance. In: Advances in Seed Priming. Berlin/Heidelberg, Germany: Springer; 2018. pp. 127-145

[51] Pescador L, Fernandez I, Pozo MJ, Romero-Puertas MC, Pieterse C, Martínez-MedinaA.Nitricoxidesignalling in roots is required for MYB72-dependent systemic resistance induced by *Trichoderma volatile* compounds in Arabidopsis. Journal of Experimental Botany. 2022;**73**:584-595. DOI: 10.1093/jxb/erab294

[52] Panchalingam H, Powell D, Adra C, Foster K, Tomlin R, Quigley BL, et al. Assessing the various antagonistic mechanisms of *Trichoderma* strains against the brown root rot pathogen *Pyrrhoderma noxium* infecting heritage fig trees. Journal of Fungi. 2022;**8**:1105. DOI: 10.3390/jof8101105

[53] Risoli S, Cotrozzi L, Sarrocco S, Nuzzaci M, Pellegrini E, Vitti A. *Trichoderma*-induced resistance to Botrytis cinerea in *Solanum* species: A meta-analysis. Planning Theory. 2022;**11**:180. DOI: 10.3390/plants11020180

[54] Dugassa A, Alemu T, Woldeha-wariat Y. In-vitro compatibility

assay of indigenous Trichoderma and *Pseudomonas* species and their antagonistic activities against black root rot disease (*Fusarium solani*) of faba bean (*Vicia faba* L.). BMC Microbiology. 2021;**21**:115. DOI: 10.1186/s12866-021-02181-7

[55] Herrera-Téllez VI, Cruz-Olmedo AK, Plasencia J, Gavilanes-Ruíz M, Arce-Cervantes O, Hernández-León S, et al. The protective effect of *Trichoderma asperellum* on tomato plants against *Fusarium oxysporum* and *Botrytis cinerea* diseases involves inhibition of reactive oxygen species production. International Journal of Molecular Sciences. 2019;**20**:2007. DOI: 10.3390/ijms20082007

[56] Zhao L, Wang Y, Kong S. Effects of *Trichoderma asperellum* and its siderophores on endogenous auxin in Arabidopsis thaliana under iron-deficiency stress. International Microbiology. 2020;**23**:501-509

[57] Ghosh SK, Banerjee S, Sengupta C. Siderophore production by antagonistic fungi. Journal of Biopesticides. 2017;**10**:105-112

[58] Chowdappa S, Jagannath S, Konappa N, Udayashankar AC, Jogaiah S. Detection and characterization of antibacterial siderophores secreted by endophytic fungi from cymbidium aloifolium. Biomolecules. 2020;**10**:1412

[59] Oszust K, Cybulska J, Frąc M. How do *Trichoderma* genus fungi win a nutritional competition battle against soft fruit pathogens? A report on niche overlap nutritional potentiates. International Journal of Molecular Sciences. 2020;**21**:4235

[60] Saravanakumar K, Wang MH. Isolation and molecular identification of *Trichoderma* species from wetland soil and their antagonistic activity against phytopathogens. Physiological and Molecular Plant Pathology. 2020;**109**:101458

[61] Bargaz A, Lyamlouli K, Chtouki M, Zeroual Y, Dhiba D. Soil microbial resources for improving fertilizers efficiency in an integrated plant nutrient management system. Frontiers in Microbiology. 2018;**9**:1606

[62] Nusaibah SA, Musa H. A review report on the mechanism of *Trichoderma* spp. as biological control agent of the basal stem Rot (BSR) disease of *Elaeis guineensis*. In: Shah MM, Sharif U, Buhari TR, editors. Trichoderma the Most Widely Used Fungicide. London: IntechOpen; 2019

[63] Kachroo A, Robin GP. Systemic signaling during plant defense. Current Opinion in Plant Biology. 2013;**16**:527

[64] Gomes EV, Costa M, de Paula RG, de Azevedo RR, da Silva FL, Noronha EF, et al. The Cerato-Platanin protein Epl-1 from *Trichoderma harzianum* is involved in mycoparasitism, plant resistance induction and self-cell wall protection. Scientific Reports. 2015;**5**:17998. DOI: 10.1038/srep17998

[65] Ngo MT, Nguyen MV, Han JW, Park MS, Kim H, Choi GJ. In vitro and in vivo antifungal activity of Sorbicillinoids produced by *Trichoderma longibrachiatum*. Journal of Fungi. 2021;7:428. DOI: 10.3390/jof7060428

[66] Matas-Baca MÁ, Urías García C, Pérez-Álvarez S, Flores-Córdova MA, Escobedo-Bonilla CM, Magallanes-Tapia MA, et al. Morphological and molecular characterization of a new autochthonous *Trichoderma* sp. isolate and its biocontrol efficacy against Alternaria sp. Saudi Journal of

Biological Sciences. 2022;**29**:2620-2625. DOI: 10.1016/j.sjbs.2021.12.052

[67] Zaid R, Koren R, Kligun E, Gupta R, Leibman-Markus M, Mukherjee PK, et al. Gliotoxin, an immunosuppressive fungal metabolite, primes plant immunity: Evidence from *Trichoderma virens*-tomato interaction. MBio. 2022;**13**:e0038922. DOI: 10.1128/mbio.00389-22

[68] Zhu N, Zhou JJ, Zhang SW, Xu BL. Mechanisms of *Trichoderma longibrachiatum* T6 fermentation against Valsa mali through inhibiting its growth and reproduction, pathogenicity and gene expression. Journal of Fungi. 2022;**8**:113. DOI: 10.3390/jof8020113

[69] Bhaskar R, Xavier LSE, Udayakumaran G, Kumar DS, Venkatesh R, Nagella P. Biotic elicitors: A boon for the in-vitro production of plant secondary metabolites. Plant Cell Tissue and Organ Culture. 2021;**147**:1-18

[70] Saravanakumar K, Fan L, Fu K, Yu C, Wang M, Xia H, et al. Cellulase from *Trichoderma harzianum* interacts with roots and triggers induced systemic resistance to foliar disease in maize. Scientific Reports. 2016;**6**:35543. DOI: 10.1038/srep35543

[71] El-Hasan A, Walker F, Klaiber I, Schöne J, Pfannstiel J, Voegele RT. New approaches to manage Asian soybean rust (*Phakopsora pachyrhizi*) using *Trichoderma* spp. or their antifungal secondary metabolites. Meta. 2022;**12**:507. DOI: 10.3390/metabo12060507

[72] Mukhopadhyay R, Kumar D. *Trichoderma*: A beneficial antifungal agent and insights into its mechanism of biocontrol potential. Egyptian Journal of Biological Pest Control. 2020;**30**(1):133. DOI: 10.1186/s41938-020-00333-x

[73] Alfiky A, Weisskopf L. Deciphering *Trichoderma*-plant-pathogen interactions for better development of biocontrol applications. Journal of Fungi. 2021;**7**:61. DOI: 10.3390/jof7010061

[74] Nawrocka J, Małolepsza U, Szymczak K, Szczech M. Involvement of metabolic components, volatile compounds, PR proteins, and mechanical strengthening in multilayer protection of cucumber plants against *Rhizoctonia solani* activated by *Trichoderma atroviride* TRS25. Protoplasma. 2018;**255**:359-373. DOI: 10.1007/s00709-017-1157-1

[75] Thambugala KM, Daranagama DA, Phillips A, Kannangara SD, Promputtha I. Fungi vs. fungi in biocontrol: An overview of fungal antagonists applied against fungal plant pathogens. Frontiers in Cellular and Infection Microbiology. 2020;**10**:604923. DOI: 10.3389/fcimb.2020.604923

[76] Kong WL, Ni H, Wang WY, Wu XQ. Antifungal effects of volatile organic compounds produced by *Trichoderma koningiopsis* T2 against *Verticillium dahliae*. Frontiers in Microbiology. 2022;**13**:1013468. DOI: 10.3389/fmicb.2022.1013468

[77] Li M, Ren Y, He C, Yao J, Wei M, He X. Complementary effects of dark septate endophytes and *Trichoderma* strains on growth and active ingredient accumulation of Astragalus mongholicus under drought stress. Journal of Fungi. 2022;**8**:920. DOI: 10.3390/jof8090920

[78] Rincón AM, Benítez T, Codón AC, Moreno-Mateos MA. Biotechnological aspects of *Trichoderma* spp. In: Rai M, Bridge PD, editors. Applied Mycology. London, UK: CAB International; 2009. pp. 216-223

[79] Sood M, Kapoor D, Kumar V, Sheteiwy MS, Ramakrishnan M,

Landi M, et al. *Trichoderma*: The "secrets" of a multitalented biocontrol agent. Plants. 2020;**9**:762

[80] Khan RAA, Najeeb S, Mao Z, Ling J, Yang Y, Li Y, et al. Bioactive secondary metabolites from *Trichoderma* spp. against phytopathogenic bacteria and root-knot nematode. Microorganisms. 2020;**8**:401. DOI: 10.3390/microorganisms8030401

[81] Esparza-Reynoso S, Ruíz-Herrera LF, Pelagio-Flores R, Macías-Rodríguez LI, Martínez-Trujillo M, López-Coria M, et al. *Trichoderma atroviride*-emitted volatiles improve growth of Arabidopsis seedlings through modulation of sucrose transport and metabolism. Plant, Cell & Environment. 2021;**44**:1961-1976

[82] Subedi S, Dhunaganaa B, Adhikarib A, Regmia V, Dhungana S. Effect of organic manures on growth and yield of cowpea in Chitwan, Nepal. Plant Physiological and Soil Chemistry. 2022;**2**(2):4-57

[83] Pandit MA, Kumar J, Gulati S, Bhandari N, Mehta P, Katyal R, et al. Major biological control strategies for plant pathogens. Pathogens. 2022;**11**(2):273

[84] Odebode AC. Control of postharvest pathogens of fruits by culture filtrate from antagonistic fungi. Journal of Plant Protector Research. 2006;**46**(1):1-5

[85] Álvarez-García S, Mayo-Prieto S, Gutiérrez S, Casquero PA. Self-inhibitory activity of *Trichoderma* soluble metabolites and their antifungal effects on *Fusarium oxysporum*. Journal of Fungi. 2020;**6**:176. DOI: 10.3390/jof6030176

[86] Zhang S, Gan Y, Liu J, Zhou J, Xu B. Optimization of the fermentation media and parameters for the bio-control potential of *Trichoderma longibrachiatum* T6 against nematodes. Frontiers in Microbiology. 2020;**11**:574601

[87] Al-Askar AA, Saber W, Ghoneem KM, Hafez EE, Ibrahim AA. Crude citric acid of *Trichoderma asperellum*: Tomato growth promotor and suppressor of *Fusarium oxysporum* f. sp. lycopersici. Plants. 2021;**10**:222. DOI: 10.3390/plants10020222

[88] Degani O, Dor S. *Trichoderma* biological control to protect sensitive maize hybrids against late wilt disease in the field. Journal of Fungi. 2021;**7**:315. DOI: 10.3390/jof7040315

[89] Zhang C, Wang W, Hu Y, Peng Z, Ren S, Xue M, et al. A novel salt-tolerant strain *Trichoderma atroviride* HN082102.1 isolated from marine habitat alleviates salt stress and diminishes cucumber root rot caused by *Fusarium oxysporum*. BMC Microbiology. 2022;**22**:67. DOI: 10.1186/s12866-022-02479-0

[90] Zhang Y, Xiao J, Yang K, Wang Y, Tian Y, Liang Z. Transcriptomic and metabonomic insights into the biocontrol mechanism of *Trichoderma asperellum* M45a against watermelon *Fusarium* wilt. PLoS One. 2022;**17**:e0272702. DOI: 10.1371/journal.pone.0272702

[91] John RP, Tyagi RD, Prévost D, Brar SK, Pouleur S, Surampalli RY. Mycoparasitic *Trichoderma viride* as a biocontrol agent against *Fusarium oxysporum* f. sp. adzuki and *Pythium arrhenomanes* and as a growth promoter of soybean. Journal of Crop Protection. 2010;**29**:1452-1459

[92] Mbarga JB, Ten Hoopen GM, Kuaté J, Adiobo A, Ngonkeu MEL, Ambang Z, et al. *Trichoderma asperellum*: A potential biocontrol agent for *Pythium myriotylum*, causal agent of cocoyam (*Xanthosoma sagittifolium*) root rot

disease in Cameroon. Crop Protection. 2012;**36**:18-22

[93] Ahmed AS, Ezziyyani M, Sánchez CP, Candela ME. Effect of chitin on biological control activity of *bacillus spp*. and *Trichoderma harzianum* against root rot disease in pepper (*Capsicum annuum*) plants. European Journal of Plant Pathology. 2003;**109**:633-637

[94] Ramezani H. Biological control of root-rot of eggplant caused by *Macrophomina phaseolina*. American-Eurasian Journal of Agricultural & Environmental Sciences. 2008;**4**:218-220

[95] Ezziyyani M, Requena ME, Egea-Gilabert C, Candela ME. Biological control of Phytophthora root rot of pepper using *Trichoderma harzianum* and *Streptomyces rochei* in combination. International Journal of Environmental Sciences. 2007;**155**:342-349

[96] Paulitz TC, Ahmad JS, Baker R. Integration of *Pythium Nunn* and *Trichoderma harzianum* isolate T-95 for the biological control of Pythium damping-off of cucumber. Plant and Soil. 1990;**121**:243-250

[97] Lewis JA, Papavizas GC. Application of *Trichoderma* and *Gliocladium* in alginate pellets for control of Rhizoctonia damping-off. Plant Pathology. 1987;**36**:438-446

[98] Howell CR. Cotton seedling preemergence damping-off incited by *Rhizopus oryzae* and *Pythium* spp. and its biological control with *Trichoderma* spp. Phytopathology. 2002;**92**:177-180

[99] El Komy MH, Saleh AA, Eranthodi A, Molan YY. Characterization of novel *Trichoderma asperellum* isolates to select effective biocontrol agents against tomato Fusarium wilt. Plant Pathology Journal. 2015;**31**:50-60

[100] Bernal-Vicente A, Ros M, Pascual JA. Increased effectiveness of the *Trichoderma harzianum* isolate T-78 against Fusarium wilt on melon plants under nursery conditions. Journal of the Science of Food and Agriculture. 2009;**89**:827-833

[101] Begum MF, Rahman MA, Alam MF. Biological control of Alternaria fruit rot of chili by *Trichoderma* species under field conditions. Mycology. 2010;**38**:113-117

[102] Amin F, Razdan VK. Potential of *Trichoderma* species as biocontrol agents of soil borne fungal propagules. Journal of Phytology. 2010;**2**:38-41

[103] Gveroska B, Ziberoski J. *Trichoderma harzianum* as a biocontrol agent against *Alternaria alternata* on tobacco. ATI-Applied Technology Innovations. 2012;**7**:67-76

[104] Rojo FG, Reynoso MM, Ferez M, Chulze SN, Torres AM. Biological control by *Trichoderma* species of *Fusarium solani* causing peanut brown root rot under field conditions. Crop Protection. 2007;**26**:549-555

[105] Freeman S, Minz D, Kolesnik I, Barbul O, Zveibil A, Maymon M, et al. *Trichoderma* biocontrol of *Colletotrichum acutatum* and *Botrytis cinerea* and survival in strawberry. European Journal of Plant Pathology. 2004;**110**:361-370

[106] Matarese F, Sarrocco S, Gruber S, Seidl-Seiboth V, Vannacci G. Biocontrol of Fusarium head blight: Interactions between *Trichoderma* and mycotoxigenic *Fusarium*. Microbiology. 2012;**158**:98-106

[107] Naeimi S, Okhovvat SM, Javan-Nikkhah M, Vágvölgyi C, Khosravi V, Kredics L. Biological control of *Rhizoctonia solani* AG1-1A, the causal agent of rice sheath blight with *Trichoderma*

strains. Phytopathologia Mediterranea. 2010;**49**:287-300

[108] Li GQ, Huang HC, Acharya SN, Erickson RS. Effectiveness of *Coniothyrium minitans* and *Trichoderma atroviride* in suppression of sclerotinia blossom blight of alfalfa. Plant Pathology. 2005;**54**:204-211

[109] Zhang H, Zhu J, Gong Z, Zhu JK. Abiotic stress responses in plants. Nature Reviews Genetics. 2022;**23**(2):104-119. DOI: 10.1038/s41576-021-00413-0

[110] Kashyap PL, Rai P, Srivastava AK, Kumar S. Trichoderma for climate resilient agriculture. World Journal of Microbiology and Biotechnology. 2017;**33**:155. DOI: 10.1007/s11274-017-2319-1

[111] Abdul-Halim AM, Shivanand P, Krishnamoorthy S, Taha H. A review on the biological properties of *Trichoderma* spp. as a prospective biocontrol agent and biofertilizer. Journal of Applied Biology & Biotechnology. 2023;**11**(5):34-46. DOI: 10.7324/JABB.2023.11504

[112] Hyakumachi M, Kubota M. Fungi as plant growth promoter and disease suppressor. In: Arora DK, editor. Fungal Biotechnology in Agricultural, Food and Environmental Application. New York: Marcel Dekker; 2003. pp. 101-110

[113] Halifu S, Deng X, Song X, Song R. Effects of two *Trichoderma* strains on plant growth, rhizosphere soil nutrients, and fungal community of *Pinus sylvestris* var. mongolica annual seedlings. Forests. 2019;**10**:758. DOI: 10.3390/f10090758

[114] Vinale F, Nigro M, Sivasithamparam K, Flematti G, Ghisalberti E, Ruocco M, et al. Harzianic acid: A novel siderophore from *Trichoderma harzianum*. FEMS Microbiology Letters. 2013;**347**:123-129

[115] Yedidia I, Srivastva AK, Kapulnik Y, Chet I. Effect of *Trichoderma harzianum* on microelement concentrations and increased growth of cucumber plants. Plant and Soil. 2001;**235**:235-242

[116] Contreras-Cornejo XA, Macías-Rodríguez L, del-Val E, Larsen J. Ecological functions of *Trichoderma* spp. and their secondary metabolites in the rhizosphere: Interactions with plants. FEMS Microbiology Ecology. 2016;**92**:1-17

[117] Tyskiewicz R, Nowak A, Ozimek E, Jaroszuk-Sciseł J. *Trichoderma*: The current status of its application in agriculture for the biocontrol of fungal phytopathogens and stimulation of plant growth. International Journal of Molecular Sciences. 2022;**23**:2329. DOI: doi. 10.3390/ijms23042329

[118] Siemering G, Ruark M, Geven A. The Value of *Trichoderma* for Crop Production. Cooperative Extension: University of Wisconsin–Extension; 2016

[119] Yedidia I, Benhamou N, Chet I. Induction of defence responses in cucumber plants (*Cucumis sativus* L.) by the biocontrol agent *Trichoderma harzianum*. Applied Environmental Microbiology. 1999;**65**:10061-11070

[120] Liu Q, Li K, Guo X, Ma L, Guo Y, Liu Z. Developmental characteristics of grapevine seedlings root border cells and their response to ρ-hydroxybenzoic acid. Plant and Soil. 2019;**443**:199-218

Chapter 6

Trichoderma: A Game Changer in the Modern Era of Plant Disease Management

Zakir Amin, Fayaz A. Mohiddin and Shazia Farooq

Abstract

Trichoderma has been found to have effectiveness against a vast range of plant diseases and can be a good alternative biocontrol strategy in the modern era of plant disease management. It has been found effective against soil borne pathogens and nematodes. *Trichoderma* has been isolated from variable soils and has multifaceted application other than disease management. *Trichoderma* enhances plant growth and development by boosting the plant's capacity to absorb nutrients, increasing systemic resistance to pest and/or pathogen attacks in the future, increasing tolerance to abiotic stresses (such as salinity, drought, and low temperatures). For instance, the stress on organic management in the modern cropping system, *Trichoderma* is a promising soil enhancer and can have handful applicability for diseases particularly those of soil borne ones. Its competitive mechanism and antagonistic approaches to compete with other pathogens makes it a good fit for future crop management strategies.

Keywords: *Trichoderma*, root rot, apple, competition, antagonist

1. Introduction

Plant disease management is one of the important strategies for a way forward towards sustainable agriculture and food security. Presently, crop output is being increased by applying herbicides, fungicides, insecticides, fertilizers, nematicides, and soil amendments [1]. Although using pesticides has some advantages, worries about pesticide residue in food are constantly growing. There is growing evidence that pesticide residue in food can have negative effects on both human health and the environment [2]. Most of these pesticides cause chronic toxicities such as genotoxicity, causing disruption in hormonal functions particularly endocrine, kidney damage, reproductive toxicity, alterations in metabolism, liver and bladder toxicity, gastrointestinal problems, etc. in addition to being potential human carcinogens, mutagens, and acetylcholinesterase inhibitors [3–5]. Taking account to a Joint UNEP and WHO research, 3 million people worldwide are poisoned by pesticides each year, killing about 200,000 people worldwide [6]. Even while high-income countries utilize large quantities of pesticides, the vast majority (95%) of pesticide poisoning incidents take place in developing countries as a result of ignorance, abuse, inappropriate management, etc. [7].

As a result, in recent years, consumers have grown increasingly concerned about the detrimental effects these synthetic fungicides have on both human well-being and the environment. Therefore, research into alternate methods of crop protection has been mandated and has gathered considerable interest from scientists all around the world. The use of biological controls by means of advantageous microorganisms has become increasingly important among these alternatives. Over millions of years of evolution, microorganisms have evolved the ability to detoxify heavy metal ions and pesticides and have become resistant to intoxicants. As a result, they have contributed to the long-term environmental benefits of restoring degraded environments to their natural state [8]. Several biological control agents (BCAs), including *Streptomyces*, *Trichoderma*, *Clonostachys*, *Bacillus*, *Pantoea*, *Pseudomonas*, *Burkholderia*, and specific yeasts, have been tested. However, the comparative effectiveness of the biocontrol agents against the chemical pesticides particularly in disease control is yet to be achieved. Therefore, the need arises for an alternative to these hazardous chemicals which could be safer to environment as well as to the humans. Although, biological control has shown a promise and effectiveness owing to their safe use and the various interactive mechanisms with plants and pathogens. However, the considerable effectiveness of the biocontrol agents with respect to the chemical pesticides particularly in disease control is yet to be achieved. There has been a debate on these biocontrol agents in terms of economics and technological perspectives worldwide leading to certain political scenarios in many countries where few protective plans to reduce the pesticide use even by up to 50% have been implemented for sustainable agriculture [9–11].

Among the various biocontrol agents like bacteria, fungus and others, the use of *Trichoderma* based formulations has grown in popularity due to its special mechanisms to compete with the various pathogens and also its growth enhancing characters. It has been thought that this particular kind of fungi is particularly advantageous for many phases of life owing to its distant sensing properties, and is quick to target and stop the spread of plant diseases, boosts plant development, and has several other features. The most significant genus of filamentous fungi in biological control techniques for managing phytopathogens is *Trichoderma* [12, 13]. Rhizospheric soils have largely been used to separate soil microorganisms, particularly the species of *Trichoderma* [14] which are considered as the probable contenders for the biocontrol aspect of plant disease management because of their critical role in inhibiting the activity of up to 80% of some economically significant plant pathogens [15] and nematodes (**Table 1**) [13]. They also play a significant role in the production of a variety of metabolites [28].

Studies across the world have revealed that *Trichoderma* species possess an outstanding ability to solubilize minerals in the soil, create hormones that promote plant growth, encourage water usage efficiency, stimulate defenses in the host, and antagonize pathogens, all of which together lead to a notable improvement in crop health [29]. It may create a variety of secondary chemicals and easily activates other fungi, creating important enzymes including chitinase, proteases, and β-1,3-glucanase that cause plant defense, systemic resistance, and fierce fight against plant diseases. In order to minimize the toxicity released by plant pathogens, it participates in a crucial detoxification process. In order to promote sustainable agriculture, it is crucial to emphasize *Trichoderma*'s significance in the treatment of plant diseases [30]. The majority of *Trichoderma* species are soil-dwelling, globally distributed molds. They take in a range of organic materials and produce a number of enzymes and secondary metabolites that could be advantageous. A few species are aggressive and dangerous to any industry that cultivates specialized edibles,

Species/strain	Biocontrol option	Reference
T. longibrachiatum T6	Pepper damping off	[16]
	Cereal cyst nematodes	[17]
T. harzianum	Pepper and Potato *Phytophthora* blight	[18–21]
	Phytopthora cactorum	[22]
T. asperellum	Apple canker	[14]
	Pythium and *Fusarium*; corn sheath blight	[23, 24]
T. harzianum	*Rhizoctonia solani; Pythium aphanidermatum*	[25]
T. viride	*Fusarium oxysporum f. sp. Adzuki; Pythium arrhenomanes*	[26]
T. virens	*Pythium ultimum; Pythophthora*	[27]

Table 1.
Trichoderma species as efficient biocontrol agents against various plant diseases.

including mushrooms. Positively, a number of species have been exploited as biological control agents because of their traits that are hazardous to plant infections. Furthermore, several *Trichoderma* species are used to promote plant development [31]. *Trichoderma* can adapt to diverse environmental circumstances and recognize the host, and the existence of a host is important for the successful colonization of the rhizosphere, plants, and soil [32]. *Trichoderma* is able to respond to the host by progressively releasing pathogenesis-related proteins (PR-Proteins), such as proteases, chitinases, and glucanases [31].

2. Morphological identification of *Trichoderma*

A unique sweet or coconut aroma is given off by several species. When examined under a microscope, conidiophores are either borne by the sparse aerial hyphae or are heavily branching, loosely or compactly tufted, frequently formed in various concentric rings, and so on. It is presumable that the typical *Trichoderma* conidiophore will have a pyramidal shape with paired branches. Phialides can have an enlarged center and can be cylindric or almost subglobose in shape. Chlamydospores could be produced by any species. Even while chlamydospores can be found inside hyphal cells, they are often unicellular subglobose and end short hyphae. Most strains of *Trichoderma* do not have a sexual stage and solely generate asexual spores. They can be identified by the growth of fleshy stromata that are light or dark brown, yellow, or orange in color. The ascomycete genus Hypocrea contains species that are the teleomorph stage of the *Trichoderma* fungus [33]. The development of fleshy stromata in hues of light or dark brown, yellow, or orange is what distinguishes them. Ascospores have two cells and tend to be green in color [34]. There are numerous species of *Trichoderma* which have been found to have application in the crop disease and pest management. Examples of those species have been documented in a wide range of environments, like that of soil from woodlands, vegetable gardens, rotting wood, cultured mushroom compost, grains of cereal, and coastal environments. Include *Trichoderma harzianum*, *T. asperellum*, *T. virens*, *T. aggressivum*, *T. longibrachiatum*, *T. reesei*, *T. citrinoviride*, *T. ghanense*, *T. hamatum*, *T. pseudokoningii*, *T. polysporum*, *T. tomentosum*, *T. atroviride* and *T. gamsii* etc. [35–37].

3. Mechanism of *Trichoderma*

As biocontrol agents against plant fungal infections, *Trichoderma* has been reported to have numerous strains. In order to combat the disease or act as a growth stimulant may be attributed to its enormously efficient mechanisms like antibiosis, parasitism, promoting host-plant resistance, and competition. *Trichoderma virens* and *Trichoderma harzianum* are active rhizosphere colonizers that produce heat-stable metabolites like ethyl acetate [38] as well as antibiotics like gliotoxin, viridin, and some enzymes that break down cell walls [39, 40]. These chemicals could contribute to the prevention of disease or the stimulation of plant growth [41]. When used as biocontrol agents, the species is particularly efficient against soilborne root diseases like root rot (*Rhizoctonia solani*), cereal take-all (*Gaeumannomyces graminis*), gray mold (*Botrytis* spp.), wilts (*Sclerotinia sclerotiorum*, *Verticillium dahliae*) and damping off (*Pythium* spp.). Conidiophores and chlamydospores are being produced on a large scale, and specialized delivery methods for these spores are being actively developed. Because of their mycoparasitic features along with their capacity to enhance plant health as well as defense, fungi from the genus *Trichoderma* can be typical biocontrol alternatives in agriculture. Consequently, it is a better plant symbiont at protecting the host from phytopathogens. *Trichoderma* works by secreting secondary metabolites and effector molecules that facilitate the advantageous interaction of *Trichoderma* with plants and confer tolerance to biotic and abiotic stimuli. Here, we go over the most recent developments in our knowledge of how this opportunist plant symbiont operates as a means of biocontrol and a plant growth booster [30]. The *Trichoderma* with its modes of action encompasses a range of procedures notably the synthesis of antibiotics, mycoparasites, competition, and systemic resistance. It also has the capability to encourage plant defense. These are the several modalities of action mentioned below [42].

3.1 Mycoparasitism

Mycoparasitism refers to the direct attack of one fungus on another, also known as direct antagonism [43]. This idea was first proposed in the work of Weindling [44], who revealed how *Trichoderma* virens hyphae could possibly parasitize *Rhizoctonia solani* hyphae to suppress citrus seedling disease. Other studies have also demonstrated *Trichoderma* species' mycoparasitism of *Pythium ultimum* and *Sclerotium rolfsii* [45]. According to Dix and Webster, mycoparasitism is a three-step, complex process that entails chemotrophic development and identification, coiling and interacting hyphae, and the production of certain lytic enzymes [43]. In order to produce infectious structures and secrete enzymes, *Trichoderma* engages in a mycoparasitic interaction with other filamentous fungi that appears to be automated by host signals [46]. The chemotropic progress of the mycoparasite's hyphae towards a target is the first noticeable contact between *Trichoderma* and its host in vitro and is probably a reaction to diffusible signals like oligochitins [47, 48]. Mycoparasite hyphae often coil around or attach to the host by forming hook-like structures when they reach it.

It was speculated decades back that mycelium permeates the host by gradually degrading its cell wall [49]. Following that, it became apparent that *Trichoderma* does, in fact, emit a sophisticated set of CWDEs (cell wall degrading enzymes) that break down cell walls and membranes [50–52]. The production of CWDEs, including chitinases, proteases, and β-1,3-glucanases, is what enables *Trichoderma* spp. to function as mycoparasitic fungi [31, 49, 53]. These results eventually contributed to the identification of the very first genes that contribute to mycoparasitism, whose expression is transcriptionally responsive to the existence of specific microbial host [50, 51]. The vast majority of mycoparasitism-related genes that have been revealed so far encode CWDEs, which are believed to operate in collaboration with secondary metabolites and are speculated to play a role in *Trichoderma*'s antagonistic ability [54]. In reality, the genus *Trichoderma* most likely descended from a distant relative that consumed either fungus or arthropods, according to a recent phylogenomic investigation of Hypocrean fungi (which included nine *Trichoderma* spp.). According to the research, entomoparasitic fungi and the monophyletic family Hypocreaceae had a last common ancestor [55].

It is interesting to take into account that signals leading to the synthesis of cell wall-degrading enzymes, the signal these enzymes produce, and the regulation of their expression in *Trichoderma* mycoparasitism exhibit significant similarities to insect pathogenicity in species of *Metarhizium*, possibly the most thoroughly noticed group of entomopathogenic fungi [50, 51, 56, 57]. Additionally, similar to *Trichoderma*, some *Metarhizium* species have recently been found to be rhizosphere competent and to form advantageous relationships with plants [58–60].

Furthermore, at least three *Trichoderma* genomes (*T. atroviride*, *T. reesei*, and *T. virens*) have had their genomes sequenced, each of which contains genes linked to the synthesis of secondary metabolites. The majority of NRPS are found in *T. virens* (28), subsequently followed by 19 alongside 16 in its two most close rivals *T. atroviride* and *Fusarium graminearum*, respectively. It is important to point out that the genetic material of *T. reesei* genome lacks half the number of secondary metabolite clusters of genes observed in *T. virens* and *T. atroviride*, which have been determined to be productive mycoparasites [61]. Genome-wide gene expression investigations were made possible by the development of genomic technology and consequently, 66 genes that were overexpressed during the start of mycoparasitism were identified by an EST-based investigation of gene expression during the interaction of *T. atroviride* with *Botrytis cinerea*/*R. solani* [62]. The majority of genes were associated with the transmission of signals, amino acid metabolic processes, post-translational processing, and fat catabolism.

Comparative transcriptome analysis revealed that before coming into contact with the host hyphae, the *Trichoderma* species *viz*., *T. virens*, *T. reesei* and *T. atroviride* responded in dramatically different ways. The gliotoxin-producing genes were expressed by *T. virens*, whereas the secondary metabolite-producing genes, β-glucanases, different proteases, and minute secreted cysteine-rich proteins were produced by *T. atroviride*. It's important to note that *T. reesei* boosted the expression of genes related to cellulose breakdown [63]. As a result, various *Trichoderma* species and even isolates employ a variety of methods, which might constitute an important aspect in the synthesis of biofungicides.

3.2 Plant interactions

Trichoderma accelerates plant development and growth on an array of crops when used in the field as well as axenic environments [64, 65]. In this regard, to help with nutrition and water uptake, the fungal mycelium secretes substances that encourage root growth [66]. *Trichoderma* species can also alter plant defense mechanisms by engaging with their various signaling pathways. It is widely acknowledged that manipulating the SA route makes it feasible for there to be a link between the fungus *T. harzianum* and the plant *Arabidopsis thaliana* because SA limits the volume of root tissue that *T. harzianum* can colonize [67]. However, fungi like *T. atroviride* have the ability to delicately alter both the SA and JA pathways to enhance resistance [68]. Secreted proteins are crucial in enabling connection between *Trichoderma* and plants, as has been demonstrated through interactions between plant-pathogens and mycorrhizae. The *Trichoderma* genomes contain hundreds of possible effector proteins, according to recent research [18, 32, 69, 70]. In a few cases, evidence has been established that confirms the involvement of effectors in the establishment of the *Trichoderma*-root interaction. In addition to effectors, *T. virens* secretes chemicals associated to secondary metabolism that aid the fungus in scavenging reactive oxygen species and hydrolyzing the plant cell wall, according to protein profiles derived from maize root tissue in interaction with the organism [70]. Using the projected secretome, researchers demonstrated differential expression of thioredoxin (tatrx2) from *T. atroviride*-derived genes with predicted activities pertaining to different families throughout the *Trichoderma*-plant interaction. Thioredoxins serve as essential effectors due to their ability to counteract the apoptotic events brought on by stress-activated signaling through mitogen-activated protein kinases (MAPK) cascades [71, 72]. The positive interaction between *T. atroviride* and *Arabidopsis* which leads in plant resistance may be mediated by tatrx2, which could play a part. It was also found that the CFEM (Common in Fungal Extracellular Membrane) proteins, which have a significant role in appressorium development in *Magnaporthe oryzae*, are expressed variably by the *T. atroviride* tacfem1 gene [73]. The genes encoding members of the LysM repeats family, tvlysm1 and talysm1 from *T. virens* and *T. atroviride*, respectively, showed differential expression as a result of the *Trichoderma-Arabidopsis* interaction. When LysM proteins bind chitin oligomers generated by the fungus, they prevent the plant from recognizing them and promote colonization [74], so that they might strengthen *Trichoderma*'s propensity to associate with plants. Finally, the *T. virens* gene tvhydii1—encoding a class II hydrophobin—plays an instrumental part in the colonization of plants [18].

The synthesis of phytohormones is an intriguing idea that might help to explain how We surveyed the *T. atroviride* genome for this review to recognize genes that may be related to the synthesis and signaling of substances like SA or JA, as well as other phytohormones. The SA, JA, Auxins, CKs, GAs, Et, and ABA pathway protein sequences in *A. thaliana* that are involved in biosynthesis and signaling were used as the seed in a BLASTP analysis against the *T. atroviride* genome database (v.2.0). The sequences with the highest similarity scores for each protein were selected using BLASTP against the genomes of *Marchantia polymorpha* and *Physcomitrella patens* in addition to this method.

3.3 Priming of plant defense responses

The molecular basis of the induced systemic resistance (ISR) brought on by *T. hamatum* in *A. thaliana* has been explored through microarrays. It was determined utilizing mutants impaired in multiple defense-related pathways that treatment with *T. hamatum* expedited the stimulation of the defensive response towards *B. cinerea*, which was perceived as an ISR-boost [75]. Similar to this, Morán-Diez et al. [76] examined *A. thaliana* gene expression alterations in response to *T. harzianum* exposure and discovered changes in genes relevant to abiotic as well as biotic stress responses, involving a number of signal transduction pathways regulated by plant hormones. Their findings are consistent with the theory that the presence of *T. harzianum* in *A. thaliana* resulted in the downregulation of genes associated to SA and JA and the upregulation of numerous genes involved in abiotic stress responses. They therefore proposed the theory that plant defenses mediated by JA and SA are less early in the encounter, allowing for root colonization. As a result, the host plant did not perceive *Trichoderma* as a threat, yet continued exposure to *Trichoderma* might activate both local and systemic defense responses. The function of some effector proteins has been thoroughly studied since they may be responsible for the activation of the plant defense response. While its *T. atroviride* ortholog (Epl1) produces systemic resistance to disease in cotton along with maize towards *Colletotrichum* spp., the *T. virens* hydrophobin-like protein Sm1 confers resistance in tomato plants from the diseases *A. solani* and *B. cinerea*. The level of expression of genes related to the JA defense mechanism or those that code for peroxidase and α-dioxygenase activity are correlated with this protection [77, 78]. Sm2 and Epl2, their paralogs, appear to be crucial for the *Trichoderma*-maize interaction in addition to Sm1 and Epl1, improving resistance and limiting the degree of leaf lesions driven by *Cochliobolus heterostrophus* [79]. Likewise, the *T. longibrachiatum* hydrophobin HYTLO1 triggers amplification of genes engaged in the SA and JA driven defense pathways, boosting protection against *B. cinerea* in tomato and pepper plant cultivars [80], TasSwo from *T. asperellum* also does the same for cucumber plants [81]. Other *Trichoderma* compounds that are not protein-based have the power to activate plant defense mechanisms. The tomato marker genes for both JA/Et and SA mediated defense pathways were expressed more frequently as a result of the secondary metabolite harzianolide from *T. harzianum* [82]. Thus, it is suggested that effector proteins and secondary metabolites are important for the development of the *Trichoderma*-plant symbiosis as well as for initiating plant defensive responses by promoting the production of SA and/or JA. However, it is conceivable that *Trichoderma* could also directly convey these signaling molecules in addition to the release of protein effectors.

3.3.1 Salicylic acid

One of the most crucial defenses against biotrophic fungus is facilitated by salicylic acid (SA), a phenolic molecule that is accountable for a variety of physiological actions in plants. It also serves as a signal molecule during plant defense. The isochorismate pathway (IC) and the phenylpropanoid pathway are the two processes through which SA is produced in plants. Both pathways start with chorismate as a precursor, which is the end product of the shikimate pathway [83]. Whereas PAL is an important enzyme that catalyzes the preliminary step in the phenylpropanoid route, converting phenylalanine into cinnamic acid, IC synthase (ICS or SID2) is an important enzyme which catalyzes the initial part of the IC pathway, transforming chorismate into isochorismate [83].

3.4 Plant growth stimulation

Numerous studies showing that *Trichoderma* spp. increase nutrient solubility and root capacity when applied to soil, seeds, or plant surfaces support the effectiveness of these organisms as biofertilizers. *Trichoderma* has a positive impact on plants, which can be attributed to the fungus' ability to modify root architecture and/or its synthesis of siderophores and organic acids, which boost nutrient availability [84]. Recent research has revealed that pathogenic fungus-plant interactions are significantly influenced by phytohormones produced by fungi. However, there is little evidence on their involvement in advantageous interactions between fungus and plant.

3.4.1 Auxins

Auxins are a series of indole-derived substances that, among other biological functions, control root initiation, cell division, and elongation in plants. Auxin-producing fungi, that mainly affect the germination of spores and elongation of cells, have been illustrated to encompass certain phytopathogenic fungi, notably species of *Fusarium*, *Rhizoctonia*, and *Colletotrichum gloeosporioides* [85, 86]. Auxins appear to have a physiological function in *Phycomyces blakesleeanus* as evidenced by the fact that exogenous delivery of auxins stimulated the growth of the fungus and the sporangiophore developing zone [87]. Tryptophan (Trp)-independent and Trp-dependent pathways make up the majority of the very complicated biosynthesis of auxins in plants [88]. In order to synthesize IAA, four Trp-dependent routes have been proposed: indole-3 pyruvic acid (IPA), indole-3 acetaldoxime (IAOx) pathway, indole-3 acetamide (IAM), and the YUCCA (YUC) pathways [88, 89]. Through the application of the IAM pathway in species of *Fusarium* and *C. gloeosporoides* and the IPA pathway in *Ustilago* and *Rhizoctonia* species, the fungi have been shown to produce auxin [86, 90]. Auxin production is perceived to guide plant root development since *T. virens* is capable of producing at least two auxin-related substances, commonly indole-3-acetic acid (IAA) and indole-3-acetaldehyde (IAAld), by means of a Trp-dependent mechanism [91]. The speculated connection among the auxins generated by the fungus and the regulation of the architecture of roots, however, is still up for question [92, 93], because volatile organic compounds produced by various *Trichoderma* strains also stimulate changes in root architecture [93]. In fact, there are evidences which suggest that the volatile 6-pentyl-2H-pyran-2-one (6-PP) generated by *T. atroviride* alters the architecture of the roots of *Arabidopsis*, resulting in an increase in plant biomass [66].

3.4.2 Cytokinins

Plant hormones known as cytokinins play functions related to nutrition balance, stress tolerance, root and shoot division, and cell differentiation [94]. CKs play a central role in the development of defensive mechanisms and plant-microbe interactions. Exogenous administration of CKs to *A. thaliana* renders the infectious agent *Hyaloperonospora arabidopsidis* resilient to them. Because plants do not activate their defenses in response to cytokinin treatment in the absence of a pathogen challenge [95], cytokinins may also serve as priming agents. Multiple investigations on the priming of advantageous fungi reveal that different *Trichoderma* species can activate plant defenses [96, 97]. However, CKs encourage hyphal branching in ectomycorrhizal mycelia [98]; and aid *M. oryzae* in tolerating oxidative stress [10, 99]. Isopentenyl transferase (IPT) or tRNA-IPT, which uses the substrate dimethylallyl-diphosphate (DAMPP) in the first phase of cytokinin biosynthesis in plants, produces N6-(2-isopentenyl) adenosine-5′-triphosphate and -diphosphate ribonucleotides (iPRTP and iPRDP, respectively). Both products are afterwards transformed into trans-zeatin ribonucleotides (tZRTP/tZRDP) by cytochrome P450 monooxygenases (CYP735A1 and CYP735A2). The LONELY GUY (LOG) family of enzymes is used to create the active CK forms in the end [100]. The probable IPT and LOG genes have been described in *M. oryzae* [99] and *Claviceps purpurea* [101], and these fungi may produce CKs. At least three receptors, including histidine kinase 2 (AHK2), histidine kinase 3 (AHK3), and histidine kinase 4 (AHK4/CRE1/WOL), are implicated in plant CKs signaling in *A. thaliana*.

DOI: http://dx.doi.org/10.5772/intechopen.1003126

3.4.3 Gibberellins

Gibberellins are the plant hormones that promote growth by breaking down the DELLA proteins that inhibit it [102]. Since its discovery in *Gibberella fujikuroi*, a fungus that causes stupid seedling disease and is where the term "gibberellins" originated, in the 1920s, they have been well-known among fungi. Although many fungi, including *Neurospora crassa*, produce these hormones, the biological function of GA in fungi is still not entirely understood [86]. Ent-kaurene is formed by the following processes: (i) biosynthesis from geranyl geranyl diphosphate (GGDP); (ii) cytochrome P450 monoxygenase converts ent-kaurene to GA12; and (iii) cytoplasmic synthesis of GA19 and GA20. There are three stages in the creation of GAs in plants. Salazar-Cerezo and colleagues [103] recently analyzed the parallels and convergence in the GAs biosynthesis pathway between plants, fungi, and bacteria [103]. In contrast to plants, fungi have clusters of genes that are involved in the biosynthesis of GAs, including three cytochrome P450 monoxygenases (P450-1, P450-2, and P450-3), one desaturase (2-oxoglutarate-dependent dioxygenase, or DES), one ent-copalyl diphosphate synthase, and one GGDP synthase gene (ggs2). GAs are less researched in plant-microbe interactions, but their potential significance could make for a fascinating subject for study. The JA signaling repressor JAZ1 is known to be released by the breaking down of DELLA proteins via GAs signaling, which increases SA signaling and increases biotroph resistance while lowering the expression of genes that respond to JA [104, 105]. Salazar-Cerezo and colleagues claim that while GAs as secondary metabolites are not necessary for fungal development or growth, they may have been helpful for survival in their ecological niche because they are primarily produced when the environment is hostile to the fungi [103]. By raising the concentration of GAs in the plant, *Trichoderma* could promote plant growth and increase plant biomass.

3.4.4 Ethylene

A gaseous phytohormone called ethylene is involved in the germination of seeds, fruit ripening, and senescence of plants [106]. By influencing simultaneously the SA and JA pathways, ethylene also contributes to plant immunity [105]. Sadenosyl-L-Methionine (S-AdoMet) and 1-aminocyclopropane-1-carboxylic acid (ACC) serve as the precursors for the manufacture of ethylene in plants. The key enzymes that catalyze this route are S-AdoMet synthetase (SAM), ACC synthase (ACS), and ACC oxidase (ACO) [107]. Numerous fungi, particularly *Penicillium digitatum*, *B. cinerea*, as well as *F. oxysporum*, have been reported to generate ethylene [85, 86]. Et has an impact on hyphal development and spore germination in fungus. It is known that *B. cinerea* utilizes methionine as a precursor of α-keto-methylthiobutyric acid (KIMBA), which in turn produces ethylene [108]. Six genes associated with Ethylene biosynthesis have been identified in the *T. atroviride* genome that were previously unknown to exist in *B. cinerea*, *T. asperellum*, *P. patens*, or *Pseudomonas* sp. [109]. The genes encoding an S-adenosylmethionine synthetase, a 1-aminocyclopropane-1-carboxylate synthase, and a member of the Iron/ascorbate family of oxidoreductases are thought to be involved in Et biosynthesis [107]. This fungus may have a functional ACC synthase pathway that produces Ethylene. Ethylene affects the physiology of the plant directly and also has a connection to nutritional stress. The amount of ethylene increases when the plant is subjected to nutritional limitation, which improves auxin sensitivity and ethylene sensing and increases stress tolerance [110]. Enhanced biotic and abiotic stress resistance is one advantage of the plant's interaction with *Trichoderma* [65]. Given that *T. atroviride* genome contains genes for Et biosynthesis, it is possible that the fungus's production of this chemical contributes to the plant's ability to withstand a variety of stresses.

3.4.5 Abscisic acid

A plant hormones called abscisic acid is a compound that modulates the dormancy of seeds and development, stomatal aperture, and enables plants withstand abiotic conditions including drought and excessive salinity [111]. ABA can have either an advantageous or detrimental effect on defense in plant-pathogen interactions [112]. For instance, ABA has a favorable effect when interacting with *Alternaria brassicicola* and a negative effect when engaging with *Pseudomonas syringae* [112]. For instance, ABA has a favorable effect when interacting with Alternaria brassicicola and a negative effect when engaging with *P. syringae*. Although the molecular specifics are still poorly known, according to Flors and collaborators [113]; ABA can also impact SA-JA cross talk, controlling basal defenses. Numerous fungi have been shown to produce ABA [86]. As it has been confirmed to be active in the *M. oryzae*-rice [114] interaction and that accumulates during the early phases of the *Ustilago maydis*-maize interaction, the primary function speculated for this fungus phytohormone is as a virulence factor stimulating plant infection [115] the primary function speculated for this fungus phytohormone is as a virulence factor stimulating plant infection. *B. cinerea* has a biosynthetic route from isopentenyl diphosphate that has been described [116], It involves the genes bcaba1, bcaba2, bcaba3, and bcaba4 [117], which are homologs to the *Arabidopsis* ataba1, ataba2, ataba3, and ataba4 genes that catalyze the final steps of ABA production [118]. *T. virens* and *T. atroviride* are capable of modulating ABA in plants, as shown by Contreras-Cornejo and colleagues' [91] discovery that strains of both species may affect stomatal aperture of *A. thaliana* and leaf transpiration by activating an ABA receptor. *Trichoderma* might be able to assist plants in reducing the negative impacts of environmental stress through yet another pathway like this. Additionally, in order to modulate stomatal aperture in *Arabidopsis* and increase the plant's resistance to water stress, ABA precursor from *Trichoderma* may be helpful.

4. Conclusion

Although increasing crop productivity, the indiscriminate use of chemical pesticides has created environment related problems. The microbial application having role in green technologies may offer alternative to chemical pesticides. *Trichoderma* evolved as a versatile microorganism to address environmental issues. The researchers have realized the usefulness of *Trichoderma* species but general and marginal farmers have remained unaware about its effectiveness. An awareness among farmers along with the identification of indigenous isolates of *Trichoderma* species is the need of an hour.

Conflict of interest

The authors declare no conflict of interest.

Author details

Zakir Amin[1], Fayaz A. Mohiddin[2]* and Shazia Farooq[1]

1 Division of Plant Pathology, Faculty of Agriculture, Sher-e-Kashmir University of Agricultural Sciences and Technology of Kashmir, J&K, India

2 Section of Plant Pathology, MRCFC, Khudwani, SKUAST-Kashmir, India

*Address all correspondence to: famohiddin@rediffmail.com

References

[1] Wahab S et al. Advancement and new trends in analysis of pesticide residues in food: A comprehensive review. Plants. 2022;**11**(9):1106

[2] Malalgoda M, Simsek S. Pesticide residue in grain-based food: Effects on health, grain quality, and chemical properties of biomacromolecules. Cereal Chemistry. 2021;**98**(1):8-16

[3] Huang Y et al. Determination of multi-pesticide residues in green tea with a modified QuEChERS protocol coupled to HPLC-MS/MS. Food Chemistry. 2019;**275**:255-264

[4] Ramakrishnan B et al. Local applications but global implications: Can pesticides drive microorganisms to develop antimicrobial resistance? Science of the Total Environment. 2019;**654**:177-189

[5] Zhang M, Zeiss MR, Geng S. Agricultural pesticide use and food safety: California's model. Journal of Integrative Agriculture. 2015;**14**(11):2340-2357

[6] Meftaul IM et al. Pesticides in the urban environment: A potential threat that knocks at the door. Science of the Total Environment. 2020;**711**:134612

[7] Parven A et al. Human health risk assessment through quantitative screening of insecticide residues in two green beans to ensure food safety. Journal of Food Composition and Analysis. 2021;**103**:104121

[8] Tarfeen N et al. Microbial remediation: A promising tool for reclamation of contaminated sites with special emphasis on heavy metal and pesticide pollution: A review. Processes. 2022;**10**(7):1358

[9] O'Brien PA. Biological control of plant diseases. Australasian Plant Pathology. 2017;**46**:293-304

[10] Barratt B et al. The status of biological control and recommendations for improving uptake for the future. BioControl. 2018;**63**:155-167

[11] Macfadyen S et al. Reducing insecticide use in broad-acre grains production: An Australian study. PLoS One. 2014;**9**(2):e89119

[12] Mohiddin F et al. Why Trichoderma is considered super hero (super fungus) against the evil parasites? Plant Pathology Journal (Faisalabad). 2010;**9**(3):92-102

[13] Poveda J, Abril-Urias P, Escobar C. Biological control of plant-parasitic nematodes by filamentous fungi inducers of resistance: Trichoderma, mycorrhizal and endophytic fungi. Frontiers in Microbiology. 2020;**11**:992

[14] Ruangwong O-U et al. Biocontrol mechanisms of *Trichoderma koningiopsis* PSU3-2 against postharvest anthracnose of chili pepper. Journal of Fungi. 2021;**7**(4):276

[15] Sánchez-Montesinos B et al. Role of *Trichoderma aggressivum* f. *europaeum* as plant-growth promoter in horticulture. Agronomy. 2020;**10**(7):1004

[16] Girma A. In vitro biocontrol evaluation of some selected Trichoderma strains against the root pathogen *Fusarium oxysporum* of hot pepper (*Capsicum annum* L.) in Bure Woreda, Ethiopia. International Journal of Microbiology. 2022;**2022**:8. Article ID: 1664116. DOI: 10.1155/2022/1664116

[17] Zhu N et al. Mechanisms of *Trichoderma longibrachiatum* T6 fermentation against Valsa mali through inhibiting its growth and reproduction, pathogenicity and gene expression. Journal of Fungi. 2022;**8**(2):113

[18] Guzmán-Guzmán P et al. Identification of effector-like proteins in *Trichoderma* spp. and role of a hydrophobin in the plant-fungus interaction and mycoparasitism. BMC Genetics. 2017;**18**(1):1-20

[19] Kappel L et al. Chitin and chitosan remodeling defines vegetative development and Trichoderma biocontrol. PLoS Pathogens. 2020;**16**(2):e1008320

[20] Mahmoud GA-E et al. Controlling *Alternaria cerealis* MT808477 tomato phytopathogen by *Trichoderma harzianum* and tracking the plant physiological changes. Plants. 2021;**10**(9):1846

[21] Liu H et al. Approaches for the establishment of optimized co-culture system of multiple Trichoderma strains for culture metabolites highly effective in cucumber growth promotion. Frontiers in Microbiology. 2022;**13**:1020077

[22] Smith V, Wilcox W, Harman G. Potential for biological control of phytophthora root and crown rots of apple by Trichoderma and Gliocladium spp. Phytopathology. 1990;**80**(9):880-885

[23] Guo Y et al. Trichoderma species differ in their volatile profiles and in antagonism toward ectomycorrhiza Laccaria bicolor. Frontiers in Microbiology. 2019;**10**:891

[24] Intana W et al. *Trichoderma asperelloides* PSU-P1 induced expression of pathogenesis-related protein genes against gummy stem blight of muskmelon (*Cucumis melo*) in field evaluation. Journal of Fungi. 2022;**8**(2):156

[25] Sivan A, Chet I. Degradation of fungal cell walls by lytic enzymes of *Trichoderma harzianum*. Microbiology. 1989;**135**(3):675-682

[26] John RP et al. Mycoparasitic *Trichoderma viride* as a biocontrol agent against *Fusarium oxysporum* f. sp. *adzuki* and *Pythium arrhenomanes* and as a growth promoter of soybean. Crop Protection. 2010;**29**(12):1452-1459

[27] Howell CR, Stipanovic RD. Gliovirin, a new antibiotic from *Gliocladium virens*, and its role in the biological control of *Pythium ultimum*. Canadian Journal of Microbiology. 1983;**29**(3):321-324

[28] Mohiddin FA et al. Phylogeny and optimization of *Trichoderma harzianum* for chitinase production: Evaluation of their antifungal behaviour against the prominent soil borne phyto-pathogens of temperate India. Microorganisms. 2021;**9**(9):1962

[29] Khan MR, Mohiddin FA. Trichoderma: Its multifarious utility in crop improvement. In: Prasad R, Singh Gill S, Tuteja N, editors. New and Future Developments in Microbial Biotechnology and Bioengineering. Crop Improvement through Microbial Biotechnology. 1st ed. Amsterdam, Netherlands: Elsevier, B.V.; 2018, pp. 263-291. DOI: 10.1016/B978-0-444-63987-5.00013-X

[30] Al-Ani LKT. Trichoderma: Beneficial role in sustainable agriculture by plant disease management. In: Egamberdieva D, Ahmad P, editors. Plant Microbiome: Stress Response. Microorganisms for sustainability. Vol. 5. Singapore: Springer; 2018. DOI: 10.1007/978-981-10-5514-0_5

[31] Harman GE et al. Trichoderma species—Opportunistic, avirulent plant symbionts. Nature Reviews Microbiology. 2004;**2**(1):43-56

[32] Mendoza-Mendoza A et al. Molecular dialogues between Trichoderma and roots: Role of the fungal secretome. Fungal Biology Reviews. 2018;**32**(2):62-85

[33] Kirk PM, Cannon PF, Minter DW, et al. Ainsworth and Bisbys Dictionary of the Fungi. 10th ed. Wallingford: CABI; 2008. p. 771

[34] Samuels GJ, Samson RA. Hypocreales of the Southeastern United States: An Identification Guide. Biodiversity Series No. 4, CBS. Utrecht: Centraalbureau voor Schimmelcultures; 2006

[35] Kubicek CP et al. Genetic and metabolic diversity of Trichoderma: A case study on south-east Asian isolates. Fungal Genetics and Biology. 2003;**38**(3):310-319

[36] Błaszczyk L et al. Species diversity of Trichoderma in Poland. Journal of Applied Genetics. 2011;**52**:233-243

[37] Sariah M et al. Quantification and characterisation of Trichoderma spp. from different ecosystems. Mycopathologia. 2005;**159**:113-117

[38] Claydown K, Emerson O, Sauthwell R. The isolation of a toxic substance from the culture filtrate of Trichoderma. Phytopathology. 1987;**36**(1068):7

[39] Larito P, Webster J, Lomas N. Trichoderma viride produce gliotoxin and viridin. Transactions of British Mycological Society. 1976;**47**:535

[40] Bell D, Wells H, Markham C. In vitro antagonism of Trichoderma species against six fungal plant pathogens. Phytopathology. 1982;**72**(4):379-382

[41] Khan M, Majid S, Mohiddin F. Biological control of Fusarium Wilt of chickpea through seed treatment with the commercial formulation of *Trichoderma harzianum* and/or *Pseudomonas fluorescens*. Phytopathologia Mediterranea. 2004;**43**:20-25

[42] Naher L et al. Trichoderma spp.: A biocontrol agent for sustainable management of plant diseases. Pakistan Journal of Botany. 2014;**46**(4):1489-1493

[43] Dix NJ, Webster J. Fungal Ecology. 1st ed. London: Chapman & Hall London; 1995

[44] Weindling R. *Trichoderma lignorum* as a parasite of other soil fungi. Phytopathology. 1932;**22**(8):837-845

[45] Papavizas G. Trichoderma and Gliocladium: Biology, ecology, and potential for biocontrol. Annual Review of Phytopathology. 1985;**23**(1):23-54

[46] Mukherjee P, Mukhopadhyay A. In situ mycoparasitism of *Gliocladium virens* on *Rhizoctonia solani*. Indian Phytopathology. 1995;**48**(1):101-102

[47] Chet I, Harman G, Baker R. Trichoderma hamatum: Its hyphal interactions with *Rhizoctonia solani* and Pythium spp. Microbial Ecology. 1981;**7**:29-38

[48] Cortes C et al. The expression of genes involved in parasitism by *Trichoderma harzianum* is triggered by a diffusible factor. Molecular and General Genetics MGG. 1998;**260**:218-225

[49] Elad Y, Barak R, Chet I. Possible role of lectins in mycoparasitism. Journal of Bacteriology. 1983;**154**(3):1431-1435

[50] Carsolio C et al. Characterization of ech-42, a *Trichoderma harzianum* endochitinase gene expressed during mycoparasitism. Proceedings of the National Academy of Sciences. 1994;**91**(23):10903-10907

[51] Geremia RA et al. Molecular characterization of the proteinase-encoding gene, prb1, related to mycoparasitism by *Trichoderma harzianum*. Molecular Microbiology. 1993;**8**(3):603-613

[52] Vázquez-Garcidueñas S, Leal-Morales CA, Herrera-Estrella A. Analysis of the β-1, 3-glucanolytic system of the biocontrol agent *Trichoderma harzianum*. Applied and Environmental Microbiology. 1998;**64**(4):1442-1446

[53] Sivan A, Chet I. The possible role of competition between *Trichoderma harzianum* and *Fusarium oxysporum* on rhizosphere colonization. Phytopathology. 1989;**79**(2):198-203

[54] Schirmböck M et al. Parallel formation and synergism of hydrolytic enzymes and peptaibol antibiotics, molecular mechanisms involved in the antagonistic action of *Trichoderma harzianum* against phytopathogenic fungi. Applied and Environmental Microbiology. 1994;**60**(12):4364-4370

[55] Druzhinina IS et al. Massive lateral transfer of genes encoding plant cell wall-degrading enzymes to the mycoparasitic fungus Trichoderma from its plant-associated hosts. PLoS Genetics. 2018;**14**(4):e1007322

[56] Freimoser FM, Hu G, Leger RJS. Variation in gene expression patterns as the insect pathogen *Metarhizium anisopliae* adapts to different host cuticles or nutrient deprivation in vitro. Microbiology. 2005;**151**(2):361-371

[57] Hu G, Leger RS. A phylogenomic approach to reconstructing the diversification of serine proteases in fungi. Journal of Evolutionary Biology. 2004;**17**(6):1204-1214

[58] Behie S, Zelisko P, Bidochka M. Endophytic insect-parasitic fungi translocate nitrogen directly from insects to plants. Science. 2012;**336**(6088):1576-1577

[59] Khan AL et al. Pure culture of *Metarhizium anisopliae* LHL07 reprograms soybean to higher growth and mitigates salt stress. World Journal of Microbiology and Biotechnology. 2012;**28**:1483-1494

[60] Sasan RK, Bidochka MJ. Antagonism of the endophytic insect pathogenic fungus *Metarhizium robertsii* against the bean plant pathogen *Fusarium solani* f. sp. *phaseoli*. Canadian Journal of Plant Pathology. 2013;**35**(3):288-293

[61] Kubicek CP et al. Comparative genome sequence analysis underscores mycoparasitism as the ancestral life style of Trichoderma. Genome Biology. 2011;**12**:1-15

[62] Seidl V et al. Transcriptomic response of the mycoparasitic fungus *Trichoderma atroviride* to the presence of a fungal prey. BMC Genomics. 2009;**10**(1):1-13

[63] Atanasova L et al. Comparative transcriptomics reveals different strategies of *Trichoderma mycoparasitism*. BMC Genomics. 2013;**14**:1-15

[64] Hermosa R et al. Plant-beneficial effects of Trichoderma and of its genes. Microbiology. 2012;**158**(1):17-25

[65] Mukherjee PK et al. Trichoderma research in the genome era. Annual Review of Phytopathology. 2013;**51**:105-129

[66] Garnica-Vergara A et al. The volatile 6-pentyl-2H-pyran-2-one from *Trichoderma atroviride* regulates *Arabidopsis thaliana* root morphogenesis via auxin signaling and ETHYLENE INSENSITIVE 2 functioning. New Phytologist. 2016;**209**(4):1496-1512

[67] Alonso-Ramírez A et al. Salicylic acid prevents *Trichoderma harzianum* from entering the vascular system of roots. Molecular Plant Pathology. 2014;**15**(8):823-831

[68] Salas-Marina MA et al. Colonization of Arabidopsis roots by *Trichoderma atroviride* promotes growth and enhances systemic disease resistance through jasmonic acid/ethylene and salicylic acid pathways. European Journal of Plant Pathology. 2011;**131**:15-26

[69] Ortíz-Castro R et al. The role of microbial signals in plant growth and development. Plant Signaling & Behavior. 2009;**4**(8):701-712

[70] Nogueira-Lopez G et al. The apoplastic secretome of Trichoderma virens during interaction with maize roots shows an inhibition of plant defence and scavenging oxidative stress secreted proteins. Frontiers in Plant Science. 2018;**9**:409

[71] Lee M-Y et al. Induction of thioredoxin is required for nodule development to reduce reactive oxygen species levels in soybean roots. Plant Physiology. 2005;**139**(4):1881-1889

[72] Rivas S et al. Retraction: CITRX thioredoxin interacts with the tomato Cf-9 resistance protein and negatively regulates defence. The EMBO Journal. 2004;**23**(10):2156-2165

[73] Kulkarni RD, Kelkar HS, Dean RA. An eight-cysteine-containing CFEM domain unique to a group of fungal membrane proteins. Trends in Biochemical Sciences. 2003;**28**(3):118-121

[74] De Jonge R, Thomma BP. Fungal LysM effectors: Extinguishers of host immunity? Trends in Microbiology. 2009;**17**(4):151-157

[75] Mathys J et al. Genome-wide characterization of ISR induced in Arabidopsis thaliana by *Trichoderma hamatum* T382 against *Botrytis cinerea* infection. Frontiers in Plant Science. 2012;**3**:108

[76] Morán-Diez E et al. Transcriptomic response of *Arabidopsis thaliana* after 24 h incubation with the biocontrol fungus *Trichoderma harzianum*. Journal of Plant Physiology. 2012;**169**(6):614-620

[77] Djonovic S et al. A proteinaceous elicitor Sm1 from the beneficial fungus *Trichoderma virens* is required for induced systemic resistance in maize. Plant Physiology. 2007;**145**(3):875-889

[78] Salas-Marina MA et al. The Epl1 and Sm1 proteins from *Trichoderma atroviride* and *Trichoderma virens* differentially modulate systemic disease resistance against different life style pathogens in *Solanum lycopersicum*. Frontiers in Plant Science. 2015;**6**:77

[79] Gaderer R et al. Sm2, a paralog of the T*richoderma cerato*-platanin elicitor Sm1, is also highly important for plant protection conferred by the fungal-root interaction of Trichoderma with maize. BMC Microbiology. 2015;**15**:1-9

[80] Ruocco M et al. Multiple roles and effects of a novel Trichoderma hydrophobin. Molecular Plant-Microbe Interactions. 2015;**28**(2):167-179

[81] Brotman Y et al. Role of swollenin, an expansin-like protein from Trichoderma,

in plant root colonization. Plant Physiology. 2008;**147**(2):779-789

[82] Cai F et al. Harzianolide, a novel plant growth regulator and systemic resistance elicitor from *Trichoderma harzianum*. Plant Physiology and Biochemistry. 2013;**73**:106-113

[83] Seyfferth C, Tsuda K. Salicylic acid signal transduction: The initiation of biosynthesis, perception and transcriptional reprogramming. Frontiers in Plant Science. 2014;**5**:697

[84] López-Bucio J, Pelagio-Flores R, Herrera-Estrella A. Trichoderma as biostimulant: Exploiting the multilevel properties of a plant beneficial fungus. Scientia Horticulturae. 2015;**196**:109-123

[85] Tudzynski B, Sharon A. Biosynthesis, biological role and application of fungal hormones. In: Osiewacz HD, editor. The Mycota X: Industrial Applications. Berlin, Heidelberg, Germany: Springer-Verlag; 2002. pp. 183-211

[86] Chanclud E, Morel JB. Plant hormones: A fungal point of view. Molecular Plant Pathology. 2016;**17**(8):1289-1297

[87] Živanović BD et al. The effect of auxin (indole-3-acetic acid) on the growth rate and tropism of the sporangiophore of *Phycomyces blakesleeanus* and identification of auxin-related genes. Protoplasma. 2018;**255**:1331-1347

[88] Mano Y, Nemoto K. The pathway of auxin biosynthesis in plants. Journal of Experimental Botany. 2012;**63**(8):2853-2872

[89] Mashiguchi K et al. The main auxin biosynthesis pathway in Arabidopsis. Proceedings of the National Academy of Sciences of the United States of America. 2011;**108**(45):18512-18517

[90] Kochar M, Upadhyay A, Srivastava S. Indole-3-acetic acid biosynthesis in the biocontrol strain *Pseudomonas fluorescens* Psd and plant growth regulation by hormone overexpression. Research in Microbiology. 2011;**162**(4):426-435

[91] Contreras-Cornejo HA et al. *Trichoderma virens*, a plant beneficial fungus, enhances biomass production and promotes lateral root growth through an auxin-dependent mechanism in Arabidopsis. Plant Physiology. 2009;**149**(3):1579-1592

[92] Hoyos-Carvajal L, Orduz S, Bissett J. Growth stimulation in bean (*Phaseolus vulgaris* L.) by Trichoderma. Biological Control. 2009;**51**(3):409-416

[93] Nieto-Jacobo MF et al. Environmental growth conditions of Trichoderma spp. affects indole acetic acid derivatives, volatile organic compounds, and plant growth promotion. Frontiers in Plant Science. 2017;**8**:102

[94] Muller B, Sheen J. Arabidopsis cytokinin signaling pathway. Science's STKE. 2007;**2007**(407):cm5-cm5

[95] Albrecht T, Argueso CT. Should I fight or should I grow now? The role of cytokinins in plant growth and immunity and in the growth–defence trade-off. Annals of Botany. 2017;**119**(5):725-735

[96] Balmer A et al. The 'prime-ome': Towards a holistic approach to priming. Trends in Plant Science. 2015;**20**(7):443-452

[97] Shoresh M, Harman GE, Mastouri F. Induced systemic resistance and plant responses to fungal biocontrol agents.

Annual Review of Phytopathology. 2010;**48**:21-43

[98] Barker SJ, Tagu D. The roles of auxins and cytokinins in mycorrhizal symbioses. Journal of Plant Growth Regulation. 2000;**19**:144-154

[99] Chanclud E et al. Cytokinin production by the rice blast fungus is a pivotal requirement for full virulence. PLoS Pathogens. 2016;**12**(2):e1005457

[100] Simm S et al. Survey of genes involved in biosynthesis, transport, and signaling of phytohormones with focus on *Solanum lycopersicum*. Bioinformatics and Biology Insights. 2016;**10**:BBI. S38425

[101] Hinsch J et al. De novo biosynthesis of cytokinins in the biotrophic fungus C laviceps purpurea. Environmental Microbiology. 2015;**17**(8):2935-2951

[102] Sun T-P. The molecular mechanism and evolution of the GA–GID1–DELLA signaling module in plants. Current Biology. 2011;**21**(9):R338-R345

[103] Salazar-Cerezo S et al. Gibberellin biosynthesis and metabolism: A convergent route for plants, fungi and bacteria. Microbiological Research. 2018;**208**:85-98

[104] Hou X et al. DELLAs modulate jasmonate signaling via competitive binding to JAZs. Developmental Cell. 2010;**19**(6):884-894

[105] Pieterse CM et al. Hormonal modulation of plant immunity. Annual Review of Cell and Developmental Biology. 2012;**28**:489-521

[106] Bleecker AB, Kende H. Ethylene: A gaseous signal molecule in plants. Annual Review of Cell and Developmental Biology. 2000;**16**(1):1-18

[107] Wang KL-C, Li H, Ecker JR. Ethylene biosynthesis and signaling networks. The Plant Cell. 2002;**14**(suppl_1):S131-S151

[108] Chagué V et al. Ethylene biosynthesis in *Botrytis cinerea*. FEMS Microbiology Ecology. 2002;**40**(2):143-149

[109] Guzmán-Guzmán P et al. Trichoderma species: Versatile plant symbionts. Phytopathology. 2019;**109**(1):6-16

[110] Iqbal N et al. Current understanding on ethylene signaling in plants: The influence of nutrient availability. Plant Physiology and Biochemistry. 2013;**73**:128-138

[111] Hauser F, Waadt R, Schroeder JI. Evolution of abscisic acid synthesis and signaling mechanisms. Current Biology. 2011;**21**(9):R346-R355

[112] Asselbergh B, De Vleesschauwer D, Höfte M. Global switches and fine-tuning—ABA modulates plant pathogen defense. Molecular Plant-Microbe Interactions. 2008;**21**(6):709-719

[113] Flors V et al. Interplay between JA, SA and ABA signalling during basal and induced resistance against *Pseudomonas syringae* and Alternaria brassicicola. The Plant Journal. 2008;**54**(1):81-92

[114] Jiang C-J et al. Abscisic acid interacts antagonistically with salicylic acid signaling pathway in rice–Magnaporthe grisea interaction. Molecular Plant-Microbe Interactions. 2010;**23**(6):791-798

[115] Morrison EN, Emery RN, Saville BJ. Phytohormone involvement in the *Ustilago maydis–Zea mays* pathosystem: Relationships between abscisic acid and cytokinin levels and strain virulence

in infected cob tissue. PLoS One. 2015;**10**(6):e0130945

[116] Inomata M et al. The biosynthetic pathway to abscisic acid via ionylideneethane in the fungus *Botrytis cinerea*. Phytochemistry. 2004;**65**(19):2667-2678

[117] Gong T et al. Sequencing and transcriptional analysis of the biosynthesis gene cluster of abscisic acid-producing *Botrytis cinerea*. International Journal of Molecular Sciences. 2014;**15**(10):17396-17410

[118] Dong T, Park Y, Hwang I. Abscisic acid: Biosynthesis, inactivation, homoeostasis and signalling. Essays in Biochemistry. 2015;**58**:29-48

Chapter 7

Trichoderma spp, Allelopathies in the Rhizosphere of Plants: For the Management of Soil Borne Pathogen, *Rhizoctonia solani*

Pranjal Kr. Kaman, Daisy Senapoty, Vithanala Shiva Sai Swaroop, Hiranya Kr. Deva Nath, Apurba Das, Pranab Dutta and Amar Bahadur

Abstract

Among the many plant diseases, those brought on by soil-borne pathogens are the ones that result in significant losses. *Rhizoctonia solani*, one of many soil-borne pathogens, has been identified as a potential culprit for yield loss due to its broad host range. Prior to the development of extremely potent and selective fungicides, chemical treatment is not a practical option. However, the dangers associated with agrochemicals are reducing their use. Scientists are becoming more interested in biological management in this situation because it is an environmentally beneficial method. Biological control is the process through which one organism controls another. Trichoderma has become one of the most important biocontrol agents currently available due to its extensive antagonistic pathways. There are 89 species in this genus, and numerous strains have been discovered to be powerful biocontrol agents for plant diseases. The species *T. viride, T. hamantum, T. koningii*, and others make up the majority of the Trichoderma biocontrol agents. Direct and indirect antagonistic mechanisms are the two categories. Mycoparasitism, antibiosis, and pathogen enzyme inactivation are examples of direct methods. Indirect mechanisms include competition for nutrients and space, the activation of plant defensive systems (such as induced systemic resistance), and others. Their antagonistic characteristics are affected by a number of variables, including pH and temperature.

Keywords: antibiosis, parasitism, *Rhizoctonia solani*, stimulants, Trichoderma

1. Introduction

Both prices are rising as a result of major losses in both quality and quantity caused by plant diseases. Since the beginning of the green revolution, controlling plant diseases and pests has become more and more dependent on the extensive use of dangerous chemicals. The cost of attaining the green revolution, however, was incalculable since agrochemicals damaged the ecosystem and upset the delicate balance between its various components. Recent research revealing that the widespread use of pesticides in the previous 10 to 15 years has had little effect in lowering the losses due to pests and diseases demonstrate that chemical use has reached a saturation point. As a result, we now need to decide whether or not supporting the use of these substances is a wise course of action. According to this viewpoint, environmentally friendly crop pest and disease management has been recognised as a substitute and essential element of sustainable agriculture. In the natural environment, microorganisms have been parasitizing both uncultivated and domesticated plants for millions of years. However, scientists have recently become interested in the phenomenon of employing bacteria for disease management, also referred to as biological control. As a biological control agent for soil-found plant pathogenic fungus, *Trichoderma* species have been investigated [1]. Several strains of *Trichoderma* considerably lessened the severity of plant diseases caused by pathogens such *Gaeumannomycesgraminis* var. *tritici*, *Sclerotium rolfsii, Phythiumaphanidermatium, Fusarium oxysporum,* and *Rhizoctonia solani* in greenhouse and field conditions. Based on findings from several studies [1–4]. Trichoderma strains have the capacity to either directly or indirectly inhibit fungal phytopathogens. Mycoparasitism, antibiosis, and pathogen enzyme inactivation are examples of direct techniques. Examples of indirect pathways include competition for resources and space, environmental changes, and the activation of plant defence mechanisms [5]. Salinity, alkalinity, nutritional insufficiency, drought, and other environmental factors all allow *Trichodema* to survive [6, 7]. This chapter will clearly explain the many mechanisms involved in *Trichoderma's* antagonistic behaviour.

Trichoderma interact with pathogenic fungi to advantageously reduce the growth of pathogens. In the past, Howell *et al.* [8] demonstrated that, in the presence of favourable growth conditions, Trichoderma rapidly filled the culture space and gradually covered the agar plate's surface, preventing the growth of the pathogenic fungus *R. solani*. Infections with hyperparasites are one of the key ways that Trichoderma species eliminate infections [9–11]. The identification of pathogens and hyperparasite signalling were greatly aided by the markedly increased gene expression of proteases linked to host recognition, oligopeptide transporters, and GPCRs during hyperparasite infection [12–14]. During the hyperparasite infection of Trichoderma in hosts, there was a considerable upregulation of the expression of genes associated with hydrolase secretion. Trichoderma's capacity to become hyperparasitic in hosts was decreased by deleting these genes.

2. Role of *Rhizoctonia solani* in causing plant diseases

Soil-borne fungus is the primary reason for the fall in crop productivity. Currently, *R. solani* is one of 1258 different fungus species that have been identified as potential crop diseases [15–17]. Plant roots and lower stems can become infected by the fungus pathogen *R. solani*. There are few options for controlling *R. solani* infections because of the variety of hosts and lack of resistant plant species. The majority of researchers are more interested in biological control among the several methods for preventing soil-borne diseases that have been suggested (**Tables 1** and **2**) [15, 23–25].

Trichoderma **spp.**	**Metabolites**	**Disease control**
T. lignorum, T. virens, T. hamatum, T. harzianum and *T. pseudokoningii (Rifai)*	Unknown inhibitory substances; Extracellular metabolites or antibiotics, or lytic enzyme action	Damping-off of bean
T. virens isolates GL3 and *GL21*	Antibiotics gliovirin and gliotoxin	Damping-off of cucumber

Table 1.
Trichoderma *spp. antagonism against diseases caused by* R. solani.

***Trichoderma* spp.**	**Reference**
T. viride, T. harzianum	[18]
T. viride	[19]
T. harzianum, *T. virens* and *T. atroviride*	[20]
Trichoderma spp. Isolates	[21]
T. asperellum	[22]

Table 2.
Uses of various Trichoderma *spp. for the control of sheath blight of rice.*

3. Different mechanisms involved in the antagonism of Trichoderma

There are broadly two different mechanisms in which *Trichoderma* inhibits the growth of phytopathogenic fungus.

1. Direct mechanism

2. In direct mechanism

3.1 Direct mechanism

In this approach, the biocontrol agent uses many mechanisms, such as mycoparasitism, antibiosis, pathogen enzyme inactivation, etc. to directly impede the growth of the phytopathogenic fungus.

3.1.1 Mycoparasitism

Mycoparasitism is another name for this biocontrol mechanism. This is an instance of another fungus parasitizing a harmful fungus. Fungi are recognised to cause the bulk of plant diseases, and it is well-known that many fungi can parasitize other fungi. The phenomenon of one fungus parasitizing another is referred to as mycoparasitism, hyperparasitism, direct parasitism, and interfungal parasitism. Coiling (**Figure 1**), hooks, and appressorium-like structures are used by *Trichoderma* species to adhere to host hyphae before secreting lytic enzymes to compromise the host cell wall [27]. Mycoparasites are fungi that may infiltrate other fungi and scavenge nutrients for their own growth by generating enzymes that breakdown cell walls. The majority of phytopathogenic fungi have an amorphous arrangement of the structural component

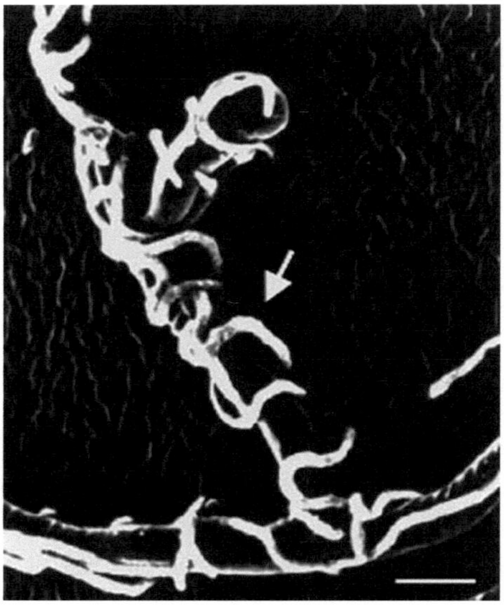

Figure 1.
Transmission electron microscopy of T. Harzianum *mycelium growing on* R. solani. *The arrow indicates the* T. Harzianum *coiling. Bar = 10 mm (source: Ref. [26]).*

chitin and the filler substance b-1,3-glucan in their cell walls. B-1,3-glucanases and chitinases have been discovered to directly affect how Trichoderma species interact with their hosts during mycoparasitism [27]. The *Trichoderma*schitinolytic system (NAGase) is composed of two different types of enzymes: chitinases and N acetyl-b-D-glucosaminidase. Despite multiple studies, it is still unclear how *Trichoderma* produces and controls chitinolytic and glucanolytic enzymes [23, 27].

3.1.2 Antibiosis

The biological control agent's capacity to suppress pathogenic organisms by secreting toxic or inhibitory substances is known as antibiosis. This occurs when a biocontrol agent and a pathogenic fungus interact, and some strains of *Trichoderma* produce low molecular weight diffusible compounds or antibiotics that inhibit the growth of other bacteria. The majority of *Trichoderma* strains release toxic compounds, both volatile and nonvolatile, that prevent fungus they have antagonised from colonising. Heptelidic acid, alamethicins, tricholin, peptaibols, 6-penthyl-pyrone, massoilactone, viridin, gliovirin, and glisoprenins are a few of the metabolites created [28]. Gliovirin can be produced by the strains of *T. virens* that are most effective for biocontrol [29]. Furthermore, the strains' performance was undoubtedly correlated with the pyrone antibiotics they produced, which are produced by the most effective isolates of *T. harzianum* against *Gaeumannomycesgraminis* var. *tritici*. Combining hydrolytic enzymes and antibiotics results in greater antagonism than using each strategy separately (**Table 3**) [29, 32].

3.1.3 Pathogen enzyme inactivation

Trichoderma degrades the toxic compounds that pathogenic bacteria secrete, which reduces their capacity to inhibit *Trichoderma* growth through the utilisation

Proteins	Encodinggene	Accession number (Gene Bank)	*Trichoderma* spp.
Peptaibol	N.A.	N.A.	*Trichoderma* spp.
Lipoxygenase	N.A.	N.A.	*T. atroviride*
Polyketide synthases (PKS)	*pks4*	N.A.	*T. atroviride*
	pks4	N.A.	*T. reesei*
	pks4	N.A.	*T. virens*
GliCCytochromeP450	*gliC*	N.A.	*T. virens*
	gliF	N.A.	*T. virens*
ɣ-glutamylcyclotransferase-likeprotein	*gliK*	N.A.	*T. virens*
Glutathione S-transferase	*gliG*	N.A.	*T. virens*
NRPSmodules	*gliP*	N.A.	*T. virens*
O-methyltransferase	*gliM*	N.A.	*T. virens*
Cytochrome P450 monooxygenases	*tri4*	FN394496	*T. arundinaceum*
Major facilitator super family transporter	*Thmfs1*	JN689385	*T. harzianum*
L-amino acidoxidase	*Th-LAAO*	GU902953	*T. harzianum*
ABC transporters	*Taabc2*	AY911669	*T. atroviride*

N.A.: Not available.

Table 3.
Trichoderma *spp. proteins associated with antagonism and involved in the synthesis of secondary metabolites deleterious to* R. solani *[30, 31].*

of heat shock proteins, ABC transporter proteins, and other mechanisms [33–35]. In order to prevent *Trichoderma* from growing and to offset the host's inhibition during contact between the two organisms, the host pathogen secretes several toxic substances. Transmembrane proteins with specific roles include ABC Transporter proteins. These proteins have two ends, one of which is referred to as a ligand binding domain and the other as an ATP binding domain. The ligand binding domain only permits a limited number of molecules to enter the cells, but the ATP binding domain eliminates any harmful substances that enter the cell. These ABC transporter proteins are triggered when some inhibitors of the host enter the cell of the biocontrol agent, and they force the toxins out of the cell before they can harm the cells [33–35].

3.2 In direct mechanism

By competing for nutrients and available space, the biocontrol agent indirectly prevents the growth of pathogenic fungus, a process known as induced systemic resistance (ISR).

3.2.1 Competition for nutrients and space

One of the most frequent causes of death for microorganisms is starvation, and fungal phytopathogens are biologically controlled by competition for few resources [36].

The majority of filamentous fungus need iron to survive, and when iron is scarce, the majority of fungi produce siderophores, low molecular weight chelators that are specific to ferric iron, to mobilise ambient iron [37]. The iron is then taken out of the ferri-siderophore complexes using specialised uptake processes. According to studies by [1], some *Trichoderma* isolates produce potent siderophores that chelate iron and inhibit the growth of other fungi (**Table 4**). *Pythium* is affected by the soil's composition and iron availability as well as Trichoderma's capacity to biocontrol it. Furthermore, *T. harzianum* competes with *Fusarium oxysporum* for nutrients and rhizosphere colonisation, preventing it from growing and boosting the efficiency of biocontrol when nutrient concentrations fall [38]. It has been discovered that *Botrytis cinerea,* the pathogenic agent that is most common throughout the pre- and post-harvest period in many countries, is particularly significant for the biocontrol of phytopathogens [39]. Because of its enormous genetic variability, the fungus can develop resistance to new chemical fungicides almost immediately after exposure [39]. The advantage of using *Trichoderma* to manage *B. cinerea* is that it allows multiple mechanisms to function at once, virtually preventing the generation of resistant strains. The most crucial of these mechanisms is nutritional competition since *B. cinerea* is so susceptible to food shortages. *Trichoderma* has a greater ability than other species to move and absorb soil nutrients (**Figure 2**).

3.2.1.1 Stimulation of plant defence mechanism induced systemic resistance (ISR)

When plants are exposed to specific signal molecules made by PGPR or other non-pathogenic fungus biocontrol agents, a phenomena known as ISR occurs. These signal molecules are circulated throughout the plant's entire system, activating its defences. *Trichoderma* strains added to the rhizosphere shield plants from a range of diseases, including viral, bacterial, and fungal pathogens, by inducing resistance mechanisms similar to the hypersensitive response (HR), systemic acquired resistance (SAR), and insensitive shock response (ISR) in plants [41, 42]. Resistance causes an increase in the concentration of metabolites and protective enzymes including phenylalanine ammoniolyase (PAL) and chalcone synthase (CHS), which are responsible for the formation of phytoalexins (HR response), chitinases, and glucanases. These comprise pathogenesis-related proteins (PR) (SAR response) and oxidative stress-responsive enzymes [43]. A range of chemicals that bacteria (elicitors) emit cause plant genes

Proteins	Encodinggene	Accession number (Gene Bank)	*Trichoderma* spp.
High-affinity glucosetransporterGtt1	*Gtt1*	AJ269534	*T. harzianum*
EndopolygalacturonaseThpg1	*Thpg1*	AM421521	*T. harzianum*
Harzianic acid	N.A.	N.A.	*T. harzianum*
Proteases	*TaPapA*	AAT09023	*T. asperellum*
	TaPapB	AAU11329	*T. asperellum*

N.A.: Not available.

Table 4.
Trichoderma spp. associated with antagonism with R. solani and involved in the competition [30, 31].

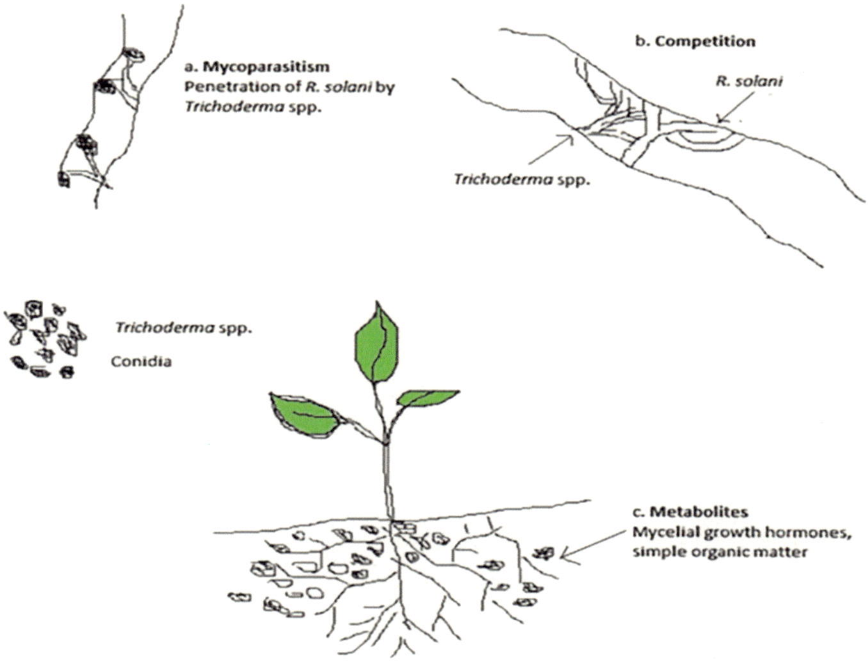

Figure 2.
Activities of Trichoderma *spp. in rhizosphere against R. solani (source: Ref. [40]).*

to respond. As a result, a pathogen is not necessarily necessary to initiate a plant's defence systems. Adding *Trichoderma* metabolites, which could trigger plant resistance. If added, *Trichoderma* elicitors, which are crucial for preventing infections in fruits after harvest, could have a similar result. When a plant is exposed to pathogens (fungi, bacteria, viruses, or nematodes), phytoalexins, antimicrobial peptides, and tiny proteins (such as thionins, defensins, hevein-like proteins, and knottin-like peptides) are created, and a number of antimicrobial proteins are upregulated [44, 45]. When *Trichoderma*, a plant pathogen, interacts with biocontrol agents, elicitors are produced. These elicitors cause plants to release something akin to PR proteins. These proteins prevent or eliminate the infection (**Tables 5** and **6**).

3.2.1.2 Factors influencing antagonistic activity of Trichoderma *species against pathogens*

The two elements that affect the antagonistic property are temperature and pH [22]; they have carried out an in vitro experiment to test the biocontrol performance of various *Trichoderma* isolates on *R. solani* at various temperatures and pH levels. At 25°C and 30°C, the three *Trichoderma* species (*T. asperellum, T. harzianum*, and *Trichoderma* spp.) showed their maximum mycelial development, however at 20°C, the growth rate was noticeably slower and they hardly colonised 1/4th of the medium surface. The pH range of 5 to 6 was the most acidic, while 6.5 to 7.0 is where mycelium grows at the slowest rate. Above these pH limits, there was either no or very little growth (0.9–1.2 cm).

Proteins	Encodinggene	Accession number (Gene Bank)	*Trichoderma* spp.
Seven-transmembrane receptorGpr1	*gpr1*	N.A.	*Trichodermaatroviride*
G-proteinone	N.A.	FD484960	*T. asperellum*
G-protein ypt3	N.A.	FD486508	*T. asperellum*
G-protein rab2	N.A.	FD485766	*T. asperellum*
α-subunit of G protein1	*tga1*	AY036905	*T. atroviride*
α-subunit of G protein3	*tga3*	AF452097	*T. atroviride*
Mitogen-activatedprotein kinases (MAPK)	*tmkA*	AY141978	*T. virens*
	tvk1	AY162318	*T. virens*
Adenylate cyclase Tac1	*tac1*	EF189190	*T. virens*
pHregulator PacC	*pacC*	N.A.	*T. virens*
pHregulator Pac1	*pac1*	EF094462	*T. harzianum*
Transcription factorThCtf1	*ctf1*	EU551672	*T. harzianum*
VELVETProteinVel1	*vel1*	N.A.	*T. virens*
XylanasetranscriptionalregulatorXyr1	*xyr1*	N.A.	*T. atrovide*

N.A.: Not available.

Table 5.
Trichoderma *spp. proteins associated with antagonism and involved in* R. solani *recognitions, signal transduction and genetic reprogramming of gene expression [30, 31].*

4. Role of Trichoderma for management of abiotic stress

In the short term, using biological agents such as water deficit-tolerant Trichoderma may be an affordable, sustainable, and eco-friendly way to lessen the effects of drought [46, 47]. However, the interactions between particular fungal isolates and plant cultivars are complicated, and our understanding of the mechanisms behind the drought tolerance that Trichoderma colonisation induces in plants is lacking. According to Hoeksema [48] coevolution involving heritable genetic variation in both plants and microbes that promote colonisation as well as abiotic stress responses may be the basis for adaptation to mutualism as a strategy of survival under harsh environmental conditions. According to Kubicek et al. [49], Trichoderma is a "environmental opportunist and generalist" that can quickly adapt to flourish as a free-living organism or a plant symbiont in novel or challenging situations. The asexual conidiospores and chlamydospores that the fungi can produce are resistant to harsh environmental conditions. Stress caused by a water scarcity considerably reduces crop quality and output. Trichoderma inoculation may help promote plant growth and lessen the harmful effects of water deficiency stress [50, 51]. It has been shown by Mastouri et al. [52, 53] that Trichoderma inoculation affects plant performance, but only in stressed conditions. According to their research, when tomato seeds were cultivated in conditions of water deprivation, the T22 inoculation enhanced germination and other growth metrics. According to

Proteins	**Encoding gene**	**Accession number (Gene Bank)**	***Trichoderma* spp.**
41-KDachitinase	*chit41*	N.A.	*T. flavus*
Chitinase1	N.A.	FD484447	*T. asperellum*
33-KDa Endochitinases	*chit33*	JK840912	*T. harzianum*
	Tv-cht1	AF395753	*T. virens*
	Tv-cht2	AF395754	*T. virens*
36-KDa Endochitinases	*chit36Y*	AF406791	*T. asperellum*
42-KDa Endochitinases	*chit42*	N.A.	*T. atroviride*
	echi42	FD485995	*T. asperellum*
	chit42	S78423	*T. harzianum*
	Tv-ech1	AF050098	*T. virens*
	Tv-ech2	AF395760	*T. virens*
46-KDa Endochitinase	*chit46*	N.A.	*T. asperellum*
Endochitinases (GH18)	*crchi1*	X80006	*T. harzianum*
N-acetyl-β-Glucosaminidases	*exc1Y*	AJ314642	*T. asperellum*
	nag1	N.A.	*T. atroviride*
	eng18B	N.A.	*T. atroviride*
	nag1	N.A.	*T. harzianum*
	Tvnag1	AF395761	*T. virens*
	Tvnag2	AF395762	*T. virens*
b-1,3-glucanases	*tag83*	EU314718	*T. asperellum*
	lam1.3	AJ002397	*T. harzianum*
29-KDab-1,3-Glucanase	N.A.	N.A.	*T. harzianum*
36-KDab-1,3-Glucanase	N.A.	N.A.	*T. harzianum*
78-KDab-1,3-Glucanase	*bgn13.1*	X84085	*T. harzianum*
b-1,6-glucanase	*bgn16.2*	N.A.	*T. harzianum*
	Tvbgn3	AF395757	*T. virens*
b-1,3-glucanase	N.A.	N.A.	*T. koningii*
	Tvbgn1	AF395755	*T. virens*
	Tvbgn2	AF395756	*T. virens*
Endo-1,3(4)-b-Glucanase	N.A.	FD486867	*T. asperellum*
Asparticproteases	*TaAsp*	EU816200	*T. asperellum*
	TaPAPA	AAT09023	*T. asperellum*
	Sa76	EF063645	*T. harzianum*
	P6281	AJ967001	*T. harzianum*
	N.A.	FD485588	*T. asperellum*

Proteins	Encoding gene	Accession number (Gene Bank)	*Trichoderma* spp.
Serineproteases	*Spm1*	FD486577	*T. asperellum*
	prb1	AAA34209	*T. harzianum*
	tvsp1	AY242844	*T. virens*
	prb1	AAA34209	*T. harzianum*

N.A.: Not available.

Table 6.
Trichoderma proteins associated with mycoparasitism of R. solani *[30, 31].*

Bashyal et al. [51], plants inoculated with Trichoderma showed higher expression levels of genes linked to photosynthesis, including those encoding for photosystem I subunits, photosystem II core complex proteins, osmotin-like proteins, and stomatal development, in comparison to non-inoculated controls. Improved physiological health is shown by plants infected with Trichoderma during water scarcity [46, 54, 55] According to reports, plants injected with Trichoderma species accumulate more metabolites, such as proline, flavonoids, and phenolics. These metabolites boost antioxidant activity and guard against photooxidative damage to photosynthetic pigments and pathways [56–58]. It is thought that plants respond to droughts by increasing the amount of abscisic acid in their tissues, which controls stomatal motions and aids in maintaining water balance. According to Martinez-Medina et al. [59], compared to non-inoculated control plants, *Trichoderma harzianum* colonised plants showed greater expression levels of the ABA-responsive marker gene involved in various defence-related pathways.

5. Benefits and limitations

5.1 Benefits

1. Trichomonia is widely used to treat post-harvest infections. *Phytophthora, Sclerotium, Fusarium*, and other fungal infections are said to respond favourably to it.

2. *Trichoderma* strains may solubilise phosphates, which really aids in plant growth in general.

3. *Trichoderma* strains are essential for bioremediation of pesticide and herbicide-contaminated soil. They are able to degrade pesticide residues made of organochlorines, organophosphates, and carbonates.

4. *Trichoderma* contains a large number of biocontrol genes. The endochitinase gene from Trichoderma has been introduced in plants, including tobacco and potato plants. These plants demonstrated enhanced infection resistance.

5. *Rhizobium, Azospirilium, Bacillus subtilis,* and other biofertilizers are compatible with *Trichoderma* as are organic manures. It works well on the seeds treated with metalaxyl and thiram, but not on seeds treated with mercurials.

5.2 Limitations

1. For up to 4 to 5 days after using Trichoderma, no chemical fungicides should be used.

2. *Trichoderma* needs adequate moisture to flourish, thus the field should not be dry.

3. *Trichoderma*-treated seeds should be dried in the shade, away from direct sunshine.

6. Conclusion

Global population pressures are increasing on scarce agricultural areas, calling for an increase in productivity. On the other hand, risky pesticide effects are restricting their use, which could temporarily lower production levels. It has been determined that using environmentally friendly pest and disease management techniques is pertinent in this situation as a component of sustainable agriculture. Because most biocontrol agents are part of natural biodiversity and are safe for the environment, biological control has therefore become increasingly important. *Trichoderma* has become quite important since it is an excellent biocontrol agent that can work against a variety of pathogenic soil fungus. Scientists are attempting to discover the genes that are responsible for its biocontrol activity in the hope of creating transgenic plants that are similar to Bt cotton using the data from recent investigations, which have yielded many hints for future research.

Author details

Pranjal Kr. Kaman[1*], Daisy Senapoty[1], Vithanala Shiva Sai Swaroop[1], Hiranya Kr. Deva Nath[2], Apurba Das[3], Pranab Dutta[4] and Amar Bahadur[5]

1 Department of Plant Pathology, Assam Agricultural University, India

2 Zonal Research Station, Shillogani, Nagaon, Assam, India

3 Department of Basic Science (Plant Pathology), College of Sericulture, India

4 College of Post Graduate Studies, Barapani, Shillong, Meghalaya, India

5 College of Agriculture, Tripura, India

*Address all correspondence to: pranjal.kaman108@gmail.com

References

[1] Chet I, Inbar J. Biological control of fungal pathogens. Appllied Biochemistry and Biotechnology. 1994;**48**:37-43

[2] Basim H, Oztork SB, Yegen O. Efficacy of a biological fungiside (planter box *Trichodermaharzianum* Rifai T-22) against seedling root rot pathogens (*Rhizoctoniasolani, Fusarium* sp.) of cotton. GAP-Environmental Symposium. ÞanlÝurfa, Turkey. 1999:137-144

[3] Chet I, Baker R. Isolation and biocontrol potential of *Trichoderma hamatum* from soil naturally suppressive to *R. Solani*. Phytopathology. 1981;**71**:286-290

[4] Sivan A, Chet I. Integrated control of media on growth and interactions between a range of soil borne glasshouse pathogens and antagonistic *fungi*. *Phytopathol*ogy. 1993;**10**:127-142

[5] Gajera H, Domadiya R, Patel S, Kapopara M, Golakiya B. Molecular mechanism of Trichoderma as bio-control agents against phytopathogen system–a review. Current Research Microbiology and Biotechnology. 2013;**1**(4):133-142

[6] Caon L, Vargas R. Threats to Soils: Global Trends and Perspectives. New York, NY, USA: Working Paper; United Nations; 2017

[7] Tandon A, Fatima T, Shukla D, Tripathi P, Srivastava S, Singh PC. Phosphate solubilization by *Trichodermakoningiopsis* (NBRI-PR5) under abiotic stress conditions. Journal of King Saud University. 2020;**32**:791-798

[8] Howell CR. Mechanisms employed by *Trichoderma* species in the biological control of plant diseases: The history and evolution of current concepts. Plant Disease. 2003;**87**:4-10

[9] Vargas WA, Mandawe JC, Kenerley CM. Multifunctional fungal plant symbionts: New tools to enhance plant growth and productivity. New Phytologist. 2011;**190**:806

[10] Woo SL, Lorito M. nExploiting the interactions between fungal antagonists, pathogens and the plant for biocontrol. In: Vurro M, Gressel J, editors. Novel Biotechnologies for Biocontrol Agent Enhancement and Management. Dordrecht, The Netherlands: Portici, NA, Italy: Springer; 2007

[11] Sharma P. Biocontrol genes from *Trichoderma* species: A review. African Journal of Biotechnology. 2011;**10**:19898-19907

[12] Suárez MB, Vizcaíno JA, Llobell A, Monte E. Characterization of genes encoding novel peptidases in the biocontrol fungus *Trichoderma harzianum* CECT 2413 using the TrichoEST functional genomics approach. Current Genetics. 2007;**51**:331-342

[13] Flores A, Chet I, Herrera-Estrella A. Improved biocontrol activity of *Trichoderma harzianum* by over-expression of the proteinase-encoding gene prb1. Current Geneics. 1997;**31**:30-37

[14] Cardoza RE, Malmierca MG, Gutiérrez S. Overexpression of erg1 gene in *Trichoderma harzianum* CECT 2413: Effect on the induction of tomato defence-related genes. Journal of Applied Microbiology. 2014;**117**:812-832

[15] Consolo VF, Monaco CI, Cordo CA, Salerno GL. Characterization of novel

Trichoderma spp. isolates as a search for effective biocontrollers of fungal diseases of economically important crops in Argentina. World Journal of Microbiology and Biotechnology. 2012;**28**(4):1389-1398

[16] Nowakowska E, Pokoj T. Changes in microbial community structure during adaptation towards polyhydroxyalkanoates production. Polish Journal of Microbiology. 2009;**58**(2):131-139

[17] Suwannarach N, Kumla J, Bussaban B, Lumyong S. Biocontrol of *Rhizoctonia solani* AG-2, the causal agent of damping-off by *Muscodorcinnamomi* CMU-Cib 461. World Journal of Microbiology and Biotechnology. 2012;**28**(11):3171-3177

[18] Montealegre J, Valderrama L, Sanchez S, Herrera R, Besoain X, Perez LM. Biological control of *Rhizoctoniasolani* in tomatoes with *Trichodermaharzianum* mutants. Electronic Journal of Biotechnology. 2010;**13**(2):1-2

[19] Da Silva JC, Torres DB, Lustosa DC, De Filippi MCC, Da Silva GB. Rice sheath blight biocontrol and growth promotion by *Trichoderma* isolates from the Amazon. Journal of Agriculture and Environmental Science. 2012;**55**:243-250

[20] Tsror L, Barak R, Sneh B. Biological control of black scurf on potato under organic management. Crop Protection. 2001;**20**(2):145-150

[21] Naeimi S, Okhovvat SM, Javan-Nikkhah M, Vágvölgyi C, Khosravi V, Kredics L. Biological control of *Rhizoctoniasolani* AG1-1A, the causal agent of rice sheath blight with *Trichoderma* strains. PhytopathologiaMediterranea. 2010;**49**(3):287-300

[22] Asad SA, Ali N, Hameed A, Khan SA, Ahmad R, Bilal M, et al. Biocontrol efficacy of different isolates of *Trichoderma* against soil borne pathogen *Rhizoctoniasolani*. Polish Journal of Microbiology. 2014;**63**(1):95-103

[23] Benitez T, Rincon AM, Limon MC, Codon AC. Biocontrol mechanisms of *Trichoderma*strains. Inernational Microbiology. 2004;**7**(4):249-260

[24] Chakraborty U, Chakraborty BN, Chakraborty AP, Dey PL. Water stress amelioration and plant growth promotion in wheat plants by osmotic stress tolerant bacteria. World Journal of Microbiology Biotechnology. 2013;**29**:789-803

[25] Vinale F, Sivasithamparam EL, Ghisalberti R, Marra SL, Woo LM, Lorito M. *Trichoderma*-plant-pathogen interactions. Soil Biology Biochemistry. 2008;**40**:1-10

[26] Almeida FBDR, Cerqueira FM, Silva RDN, Ulhoa CJ, Lima AL. Mycoparasitism studies of *Trichodermaharzianum* strains against *Rhizoctoniasolani*: Evaluation of coiling and hydrolytic enzyme production. Biotechnology Letters. 2007;**29**(8):1189-1193

[27] Kubicek CP, Mach RL, Peterbauer CK, Lorito M. *Trichoderma*: From genes to biocontrol. Journal of Plant Pathology. 2001;**83**:11-24

[28] Vey A, Hoagland RE, Butt TM. Toxic metabolites of fungal biocontrol agents. In: Butt TM, Jackson C, Magan N, editors. Fungi as Biocontrol Agents: Progress, Problems and Potential. Bristol: CAB International; 2001. pp. 311-346

[29] Howell CR. The role of antibiosis in biocontrol. In: Harman GE, Kubicek CP, editors. *Trichoderma* and *Gliocladium*.

Vol. 2. Padstow: Taylor & Francis; 1998. pp. 173-184

[30] Pozo MJ, Baek JM, Garcıa JM, Kenerley CM. Functional analysis of tvsp1, a serine protease-encoding gene in the biocontrol agent *Trichoderma virens*. Fungal Genetics and Biology. 2004;**41**(3):336-348

[31] Scherm B, Schmoll M, Balmas V, Kubicek CP, Migheli Q. Identification of potential marker genes for *Trichodermaharzianum* strains with high antagonistic potential against *Rhizoctoniasolani* by a rapid subtraction hybridization approach. Current Genetics. 2009;**55**(1):81-91

[32] Monte E. Understanding *Trichoderma*: Between biotechnology and microbial ecology. International Microbiology. 2001;**4**:1-4

[33] Bae SJ, Mohanta TK, Chung JY, Ryu MJ, Park G, Shim S, et al. *Trichoderma* metabolites as biological control agents against Phytophthora pathogens. Biological Control. 2015;**92**:128-138

[34] Lorito M, Woo SL. *Trichoderma*: A multi-purpose tool for integrated Pest management. In: *Principles of Plant-Microbe Interactions*. Portici, NA, Italy: Springer International Publishing; 2015

[35] Seidl V, Song L, Lindquist E, Gruber S, Koptchinskiy A, Sun JB, et al. Transcriptomic response of the mycoparasitic fungus *Trichodermaatroviride*to the presence of a fungal prey. BMC Genomics. 2009;**10**:567

[36] Chet I, Inbar J, Hadar I. Fungal antagonists and mycoparasites. In: Wicklow DT, Söderström B, editors. The Mycota IV: Environmental and Microbial Relationships. Berlin: Springer-Verlag; 1997. pp. 165-184

[37] Eisendle M, Oberegger H, Buttinger R. Biosynthesis and uptake of siderophores is controlled by the PacCmediated ambient-pH regulatory system in *Aspergillusnidulans*. Eukaryotic Cell. 2004;**3**:561-563

[38] Katan J, Ginzburg C, Freeman S. Tjamos EC, Papavizas GC, Cook RJ, editors. Biological Control of Plant Diseases. Progress and Challenges for the Future. New York: Plenum Press; 1992

[39] Latorre BA, Spadaro I, Rioja ME. Occurrence of resistant strains of *Botrytis cinerea* to anilinopyrimidine fungicides in table grapes in Chile. Crop Protection. 2002;**21**(10):957-961

[40] Abbas A, Jiang D, Fu Y. *Trichoderma* spp. as antagonist of *Rhizoctoniasolani*. Journal Plant Pathology and Microbiology. 2017;**8**(3):1-9

[41] Harman GE, Howell CR, Viterbo A. *Trichoderma* species-opportunistic, avirulent plant symbionts. Nature Reviews. 2004;**2**:43-56

[42] Harman GE. Overview of mechanisms and uses of *Trichoderma* spp. Phytopathology. 2006;**96**:190-194

[43] Lyon GD, Stacey G, Keen NT, editors. Plant-Microbe Interactions. St. Paul: APS Press; 1999. p. 4

[44] Epple P, Apel K, Bohlmann H. An *Arabidopsis thaliana*thionin gene is inducible via a signal transduction pathway different from that for pathogenesis-related proteins. Plant Physiology. 1995;**109**:813-820

[45] Terras FRG, Schoofs H, De Bolle MFC. Analysis of two novel classes of plant antifungal proteins from radish (*Raphanus sativus* L.) seeds. The Journal of Biological Chemistry. 1992;**267**:15301-15309

[46] Shukla N, Awasthi RP, Rawat L, Kumar J. Biochemical and physiological responses of rice (*Oryza sativa* L.) as influenced by Trichoderma harzianum under drought stress. Plant Physiology and Biochemistry. 2012;**54**:78-88

[47] Brotman Y, Landau U, Cuadros-Inostroza A, Takayuki T, Fernie AR, Chet I, et al. Trichoderma-plant root colonization: Escaping early plant defense responses and activation of the antioxidant machinery for saline stress tolerance. PLoS Pathogens. 2013;**9**(3):1003221

[48] Hoeksema JD. Ongoing coevolution in mycorrhizal interactions. New Phytologist. 2010;**187**(2):286-300

[49] Kubicek CP, Steindorff AS, ChenthamaraK,ManganielloG,HenrissatB, Zhang J, et al. Evolution and comparative genomics of the most common Trichoderma species. BMC Genomics. 2019;**20**:1-24

[50] De Palma M, D'Agostino N, Proietti S, Bertini L, Lorito M, Ruocco M, et al. Suppression subtractive hybridization analysis provides new insights into the tomato (*Solanum lycopersicum* L.) response to the plant probiotic microorganism *Trichoderma longibrachiatum* MK1. Journal of Plant Physiology. 2016;**190**:79-94

[51] Bashyal BM, Parmar P, Zaidi NW, Aggarwal R. Molecular programming of drought-challenged *Trichoderma harzianum*-bioprimed rice (*Oryza sativa* L.). Frontiers in Microbiology. 2021;**12**:655165

[52] Mastouri F, Björkman T, Harman GE. Seed treatment with *Trichoderma harzianum* alleviates biotic, abiotic, and physiological stresses in germinating seeds and seedlings. Phytopathology. 2010;**100**(11):1213-1221

[53] Mastouri F, Björkman T, Harman GE. *Trichoderma harzianum* enhances antioxidant defense of tomato seedlings and resistance to water deficit. Molecular Plant-Microbe Interactions. 2012;**25**(9):1264-1271

[54] Harman GE, Doni F, Khadka RB, Uphoff N. Endophytic strains of Trichoderma increase plants' photosynthetic capability. Journal of Applied Microbiology. 2021;**130**(2):529-546

[55] Khoshmanzar E, Aliasgharzad N, Neyshabouri MR, Khoshru B, Arzanlou M, Asgari Lajayer B. Effects of Trichoderma isolates on tomato growth and inducing its tolerance to water-deficit stress. International journal of Environmental Science and Technology. 2020;**17**:869-878

[56] Mona SA, Hashem A, Abd_Allah EF, Alqarawi AA, Soliman DWK, Wirth S, et al. Increased resistance of drought by *Trichoderma harzianum* fungal treatment correlates with increased secondary metabolites and proline content. Journal of Integrative Agriculture. 2017;**16**(8):1751-1757

[57] Scudeletti D, Crusciol CAC, Bossolani JW, Moretti LG, Momesso L, Servaz Tubana B, et al. *Trichoderma asperellum* inoculation as a tool for attenuating drought stress in sugarcane. Frontiers in Plant Science. 2021;**12**:645542

[58] Tripathi R, Keswani C, Tewari R. *Trichoderma koningii* enhances tolerance against thermal stress by regulating ROS metabolism in tomato (*Solanum lycopersicum* L.) plants. Journal of Plant Interactions. 2021;**16**(1):116-125

[59] Martínez-Medina A, Fernández I, Sánchez-Guzmán MJ, Jung SC, Pascual JA, Pozo MJ. Deciphering the hormonal signalling network behind the systemic resistance induced by *Trichoderma harzianum* in tomato. Frontiers in Plant Science. 2013;**4**:206

Chapter 8

Safeguarding Citrus: Exploring State-of-the-Art Management Strategies for Bacterial Citrus Diseases

Aditya Kukreti and Namburi Karunakar Reddy

Abstract

Bacterial diseases pose significant risks to the citrus industry, causing major economic losses and threatening global production. The most severe threats include citrus canker (*Xanthomonas citri*) which leads to visible lesions on leaves, fruit, and stems and Huanglongbing (HLB) (*Candidatus* Liberibacter spp.) which results in mottled leaves, stunted growth, and deformed, bitter fruit. Beyond these major diseases, citrus blast and citrus variegated chlorosis (CVC) are emerging concerns. Citrus blast, caused by *Pseudomonas syringae* pv. *citri*, results in leaf lesions, cankers, and defoliation, managing it involves copper-based bactericides, removing infected branches, and cultural practices to reduce spread. CVC, caused by *Xylella fastidiosa*, is a vascular disease leading to chlorosis, leaf scorch, and dieback. Management strategies for CVC include controlling insect vectors with insecticides and developing resistant citrus varieties. Integrated disease management is crucial, focusing on sustainable approaches that combine cultural practices, biological control agents, and resistant varieties. Advances in technology, such as molecular diagnostics, remote sensing, and precision agriculture, are improving early detection and monitoring. Public awareness and education are keys to encouraging growers to adopt best practices. Collaboration among researchers, growers, and policymakers remains essential to tackle the complex challenges of bacterial citrus diseases and ensure the citrus industry's sustainability.

Keywords: citrus blast, citrus variegated chlorosis, citrus canker, Huanglongbing, management

1. Introduction

Citriculture holds significant global economic importance, benefiting from the adaptability of citrus varieties to diverse regions within the intertropical range [1]. Engaging at least 140 countries, citrus cultivation yields an annual production exceeding 122 million tons of fruits. According to the United States Department of Agriculture (2023), China holds the top position as the leading citrus producer, particularly excelling in tangerines/mandarins, with an anticipated output of 26.5 million tons. Despite its economic prominence, citriculture faces threats from diseases that pose substantial risks to production due to their widespread occurrence. Bacterial diseases, such as citrus variegated chlorosis (CVC), Huanglongbing (HLB), citrus canker and citrus blast, contribute to significant economic impacts on the global citrus industry [2]. These diseases, each characterized by distinct pathogen disease cycles and reproductive strategies, exhibit broad host ranges, pathogenicity, and genomic variations, indicating horizontal gene transfer of pathogenicity factors. Monoculture practices in commercial citrus groves, where a limited number of varieties cover extensive areas, lack natural variation in resistance, further aggravating disease susceptibility [3]. To study these pathogens and to manage them, it is essential first to differentiate between parasitism and pathogenicity. While a parasite can derive nutrition from its host without necessarily causing disease, a pathogen induces disease but not always on all hosts. For instance, a Liberibacter-infected pear tree may display no symptoms [4]. Pathologists traditionally focus on pathogens affecting commercially vital crops, overlooking potential wider parasitic host ranges. Overlapping host ranges among bacterial species and the potential for insect vectors to disguise host range limitations contribute to emerging and severe disease threats. The introduction of vectors with broad host ranges, encompassing weeds or arboreal crops, can facilitate horizontal gene transfer events, intensifying recombination frequencies across species and genera. Transcontinental dispersal of pathogens and vectors further amplifies the risk of horizontal gene transfer [3]. In light of these concerns, this chapter comprehensively explores major diseases affecting global citriculture, including citrus canker, HLB, CVC, and citrus blast. The discussion covers topics ranging from the causes and symptoms of the disease to strategies for control, diagnostic methodologies, and recent advancements in alternative management approaches.

2. Etiology and symptoms of bacterial threats in citrus

Citriculture, a crucial economic activity in many countries, faces significant challenges arising from the prevalence of various diseases that pose a considerable threat to production. The frequency of occurrence and the extensive damage inflicted by these diseases have heightened concerns within the global citrus industry. Important bacterial diseases involved are mentioned in **Table 1**.

2.1 Citrus canker

Citrus canker is an important bacterial disease of citrus generally linked to two bacterial species within the *Xanthomonas* genus. *X. citri* subsp. *citri* (Xcc) is responsible for causing citrus canker A, which is identified as the most aggressive strain, posing a significant threat in Asia and South America and will be discussed here whereas *X. citri* subsp. *aurantifolii* pathotypes B and C (XauB and XauC) causing citrus canker B and C, respectively, exhibit lower aggressiveness. Citrus canker B is prevalent in Paraguay, Argentina, and Uruguay whereas citrus canker C has been restricted to Sao Paulo state in Brazil [18, 19].

The formation of biofilms by Xcc bacteria is a critical virulence factor, playing a significant role in the initial stages of infection and the colonization of hosts [20, 21]. The characteristic symptoms of asiatic citrus canker include the development of necrotic lesions with moist edges, the presence of chlorotic rings in leaves, stems, and fruits [20, 22] and heightened mitotic cell division leading to the formation of

DOI: http://dx.doi.org/10.5772/intechopen.1004879

S. No.	Diseases	Origin	Countries globally where these diseases are documented	Pathogens	Diagnostic symptoms	Resistant genotypes/ cultivars
1	Citrus canker	Southeast Asia-India, 1830–1840s.	Afghanistan, Argentina, Bangladesh, Bolivia, Brazil, Burkina Faso, Cambodia, China, Christmas Island, Cocos Islands, Comoros, Cote d'Ivoire, East Timor, Ethiopia, Fiji, Gabon, Guam, India, Indonesia, Iran, Iraq, Japan, Korea Dem. People's Republic, Korea, Republic, Laos, Madagascar, Malaysia, Maldives, Mali, Marshall Islands, Mauritius, Mayotte, Micronesia, Myanmar, Nepal, Northern Mariana Islands, Oman, Pakistan, Palau, Papua New Guinea, Paraguay, Philippines, Reunion, Senegal, Seychelles, Singapore, Solomon Islands, Somalia, Sri Lanka, Sudan, Taiwan, Tanzania, Thailand, United Arab Emirates, United States, Uruguay, Vietnam [5]	*Xanthomonas citri* subsp. *citri* pathotype A (XccA)	 Characterized by raised lesions (cankers) on susceptible citrus cultivars fruits, foliage, and young stems [3]	Valencia, Tangi, Japan; Lakeland (limequat),seed parent; Tangor Norin no. 8, Japan; Amaka [6]
2	Huanglongbing	Southwest Asia,1919	Angola, Argentina, Bangladesh, Barbados, Belize, Bhutan, Brazil, Burundi, Cambodia, Cameroon, Central African Republic, China, Colombia, Costa Rica, Cuba, Dominica, Dominican Republic, East Timor, El Salvador, Eswatini, Ethiopia, French Guiana, Guadeloupe, Guatemala, Honduras, India, Indonesia, Iran, Jamaica, Japan, Kenya, Laos, Madagascar, Malawi, Malaysia, Martinique, Mauritius, Mexico, Myanmar, Nepal, Nicaragua, Nigeria, Oman, Pakistan, Panama, Papua New Guinea, Papua New Guinea, Paraguay, Philippines, Puerto Rico, Reunion, Rwanda, Saint Helena, Saudi Arabia, Somalia, South Africa, Sri Lanka, Taiwan, Tanzania, Thailand, Trinidad and Tobago, Uganda, United States, Venezuela, Vietnam, Yemen, Zimbabwe [5]	Candidatus Liberibacter	 Appearance of asymmetric, blotchy mottles on the leaves [3]	There is no citrus varieties resistant to HLB [7]

S. No.	Diseases	Origin	Countries globally where these diseases are documented	Pathogens	Diagnostic symptoms	Resistant genotypes/cultivars
3	Citrus blast	—	Citrus blast has been reported in Turkey [8], Iran [9] and Montenegro [10], Tunisia [11]	*Pseudomonas syringae* pv. *syringae*	 Lesions initially form on the leaf petiole or wing of citrus and appear as small water-soaked or dark spots. Black pit symptoms are visible on citrus fruits [12]	Although no citrus cultivar is fully resistant to blast [12], Lemon cultivar 'Eureka' was found least susceptible [13].
4	Citrus variegated chlorosis	Argentina, 1984	Argentina, Brazil, Canada, Costa Rica, France, Iran, Israel, Italy, Mexico, Paraguay, Portugal, Puerto Rico, Spain, Taiwan, United States, Venezuela [5]	*Xylella fastidiosa* subsp. *pauca* (Xfp)	 Yellow lesions similar to zinc deficiency on the adaxial leaf side, and brown lesions on the abaxial side [3, 14, 15]	Navelina ISA 315 [16]; Murcott tangor [17].

Table 1.
Major bacterial diseases of citrus.

cankers [23, 24]. Severe asiatic citrus canker cases can lead to defoliation, dieback, and fruit drop, resulting in reduced fruit value or complete unmarketability. While most commercial citrus varieties exhibit varying degrees of susceptibility, grapefruit stands out as the most vulnerable [25]. The transmission of the disease is facilitated by rain splash and wind, with the extent of bacterial aerosol spread ranging from a few centimeters in mild rain without wind to several kilometers during tropical storms and hurricanes, with the latter conditions intensifying the outbreak [26, 27]. The severity of Asiatic citrus canker outbreaks is heightened in regions where the Asian citrus leaf miner (*Phyllocnistis citrella*) and citrus canker overlap, due to the insect-feeding activities that result in wounds that expose leaf mesophyll tissues to splashed inoculum, thus significantly increasing the likelihood of infection [28, 29]. Asiatic citrus canker exhibits high contagion, often spreading through human-mediated means such as contaminated clothing, vehicles, equipment (e.g., pruners, hedgers, mowers, ladders, harvesters), or animals, whether in commercial or residential citrus environments; the bacteria is released from hyperplastic lesions on wetted foliage and can be transferred to susceptible foliar tissues on neighboring or distant trees when coming into contact with the bacterial ooze through mechanical devices, humans, or animals [3].

2.2 Huanglongbing

Generally known as Huanglongbing (HLB) in the Americas, it initially gained recognition under different names in various regions. In South Africa, it was referred to as Greening, while in the Philippines, it was identified as Mottle Leaf. In India, the disease was termed Dieback, and in Indonesia, it was recognized as Vein Phloem Degeneration. These diverse names reflect the global distribution and regional awareness of the same underlying threat to citrus crops, showcasing the widespread impact of this disease on citrus cultivation across different continents [30, 31].

Jagoueix et al. were pioneers in employing 16S rDNA sequence comparisons to establish that this citrus pathogen belongs to the alpha subdivision of proteobacteria and categorized it under the provisional taxon Candidatus. Liberibacters constitute a closely related cluster of alphaproteobacteria within the Rhizobiaceae family. HLB is induced by three Gram-negative bacteria which were named based on the continents where they were initially identified: Candidatus *Liberibacter asiaticus* (CLas), Candidatus *Liberibacter americanus* (CLam), and Candidatus *Liberibacter africanus* (CLaf) [32–34]. Additionally, the bacteria growth in planta is systemic, phloem limited and circulative in its psyllid vectors *Diaphorina citri* and *Trioza erytreae*. Intriguingly, they display a higher degree of adaptation to their insect hosts than to plant hosts. They can spread both inter- and intracellularly across various tissues in insect hosts, including salivary glands, alimentary canal, midgut epithelium, basement membrane, Malpighian tubules, hemolymph, muscle and fat tissues, and ovaries. Remarkably, there is little to no evidence of pathogenic effects associated with this spread in insect vector [35]. The intensity of symptoms is linked to the duration of the plant's infection. Following the initial infection, a notably extended incubation period, typically lasting at least 6 months, is minimal before any observable symptoms as CLas will try to suppress the immune response by its citrus host [3, 36].

The primary identifiable symptoms of HLB disease include the yellowing of branches, the distortion of small leaves accompanied by yellow spots, fruit deformities, seed abortion, premature leaf and fruit shedding, as well as the death of sprouts. The visual characteristics of symptomatic leaves closely resemble those resulting from deficiencies in zinc, manganese, and iron, as documented by various researchers [30, 31, 37, 38]. CLas and CLam in citrus exhibit notable differences in multiplication and symptom induction. CLas demonstrates a tenfold higher multiplication in citrus compared to CLam, yet CLam induces more severe symptoms of HLB despite having a lower titer [39]. Grafting budwood for CLas transmission achieves nearly 100% efficiency, while CLam transmission through grafting is typically below 50%. Comparable data for CLaf are currently unavailable. Natural transmission by psyllids is influenced by various factors, with bacterium multiplication and temperature being crucial. The transmission of CLas by *D. citri* is affected by CLas titer in citrus, with efficient transmission occurring from citrus to citrus, while CLam transmission is less efficient [3].

2.3 Citrus blast

Citrus blast, commonly referred to as black pit, is caused by *Pseudomonas syringae* pv. *syringae* (Pss) bacterium, a Gram-negative, chemoorganotrophic bacterium belonging to the Gammaproteobacteria. The Pss bacterium is naturally present in citrus leaves and various plant species. Elevated populations of Pss strains which are associated with blast and black pit, proliferate on the surfaces of citrus leaves and twigs during extended periods of fog, rain, and low temperatures [40].

Pss bacterium infects susceptible citrus tissues, including shoots, twigs, young petioles, or fruit, entering through wounds caused by hail, wind, or abrasions [10, 41]. Lesions initially form on the leaf petiole or wing of citrus and appear as small water-soaked or dark spots. These lesions rapidly expand towards the leaf midvein, twig, and axil tissues, causing the leaves to wither, curl, and turn brown, eventually dropping, often without petioles. Severe attacks can lead to the death of entire twigs, contributing to the tree's blasted appearance, a condition sometimes mistaken for frost damage. Black pit symptoms are visible on citrus fruits, particularly lemons, which are highly susceptible. Affected fruits, especially those with physical injuries from hail or thorn and branch abrasion, exhibit light brown spots that progressively darken and become markedly sunken, forming black pits typically less than 1 cm in diameter. During postharvest storage, these pits may enlarge, resulting in fruit loss [13, 41, 42]. The occurrence of blast and black pit is prominent in regions characterized by frequent wet, cool, and windy conditions in winter and spring [40].

2.4 Citrus variegated chlorosis (CVC)

X. fastidiosa, the first plant pathogen to undergo complete genome sequencing [43], colonizes the xylem vessels of citrus plants, thus compromising plant development by blocking water and nutrients and leading to CVC [44]. Xylem colonization by Xfp initiates when the insect vector, carrying the pathogen, feeds on the xylem sap of a healthy tree. This colonization is facilitated by enzymes that degrade xylem pit membranes, preventing the obstruction of Xf movement [45]. Ultimately, Xfp colonization leads to the blockage of sap flow [46, 47], and infected trees exhibit characteristic disease symptoms first on leaves and later on fruits.

Although *X. fastidiosa* can colonize a broad spectrum of hosts, encompassing over 500 plant species, the disease itself is constrained to affecting sweet oranges (*Citrus sinensis*), tangerines, and their hybrids. Therefore, the range of host species susceptible to infection by Xfp is delimited by the phytopathogen's genome [3, 48]. Xfp can be transmitted naturally through root grafts [49] or by sharpshooters [3]. The vectors of CVC are insects that feed on citrus xylem vessels, specifically leafhoppers of the Cicadellidae family, such as *Bucephalogonia xanthophis* and *Macugonalia leucomelas*, which are particularly effective in transmitting the pathogen. Additionally, transmission can occur through spittlebugs of the Aphrophoridae family [50, 51].

CVC affected leaves display small spots on both surfaces, progressing to orange gum in the centre of the lower surface spot, with yellow spots coalescing and turning necrotic in advanced stages. Highly affected leaves often abscise, leaving tufts of small leaves on affected branches resembling zinc deficiency. Fruits on affected branches mature earlier, are smaller and harder, with increased sugar content but higher acidity compared to fruits from asymptomatic branches of infected or healthy trees [52]. Smaller fruit size leads to direct losses, requiring more affected fruits to fill a standard commercial box. The harder fruits impede juice production, and their reduced size makes them less appealing to consumers in fresh markets. The progression of CVC is more rapid in younger trees and those exposed to seasonal water deficits. Under such conditions, tree development is significantly impacted by the disease, resulting in a substantial decrease in fruit production, reaching up to 75% compared to healthy or asymptomatic trees [53].

3. Detection and management strategies for tackling diseases

3.1 Citrus canker

Despite the economic significance of citrus canker, there has been a lack of effective and environmentally friendly treatments developed to date. The approach to controlling Asiatic citrus canker depends on whether the disease is newly introduced or if it has become endemic. In instances where the disease is recently introduced with low incidence, efforts are often directed towards eradication to restore the region to a disease-free state, aiming to avoid the expenses associated with quarantines and trade restrictions. The eradication process involves the prompt removal of infected plants at a rate faster than the rate of new infections, minimizing the chances of ongoing pathogen spread and hastening the extinction of the disease. However, practical implementation is intricate, as newly infected plants may not be readily recognizable due to a common several-week incubation period before disease expression. Subclinical infections, where plants are infected but not displaying symptoms, can persist for 10–14 days for Asiatic citrus canker, making timely detection challenging. Detection in residential areas has been estimated to take 108 days post-infection, allowing ample time for polycyclic reproduction, and spread. Moreover, diseased plants are not always grouped conveniently for eradication, often presenting a diffuse pattern within a healthy population. Achieving eradication, therefore, requires the removal of not only recognized infections but also asymptomatic yet potentially latently infected plants within large radii. These eradication programs, aimed at Asiatic citrus canker epidemics, have occasionally faced resistance from commercial growers and residential homeowners due to the removal of seemingly healthy trees, impacting the social tolerance threshold and creating a challenge in garnering support for the benefit of the commercial citrus industry [3].

In regions where Asiatic citrus canker is endemic, effective mitigation involves the integration of suitable cultural practices, including the implementation of windbreaks and control measures against leaf miners. Regular applications of copper sprays, such as copper hydroxide and copper oxychloride, are recommended [54].

Notably, grapefruit exhibits high susceptibility to Asiatic citrus canker, necessitating more frequent sprays for acceptable control compared to less susceptible varieties like oranges or mandarins. Antibiotics like gentamicin, kasugamycin, and streptomycin show efficacy in canker control when properly formulated, although concerns about bacterial resistance and potential health issues for humans and animals arise with widespread antibiotic use in agriculture [55, 56]. Addressing the challenge of resistant Xcc populations could involve the implementation of strategies designed to mitigate resistance to copper bactericides, ultimately enhancing their efficacy and minimizing the need for frequent application [20, 57]. One viable alternative is the adoption of the tree row volume (TRV) methodology, initially proposed by Byers et al., which optimizes spray volume and copper application rates based on estimated canopy volume in a specific area [58]. Previous research on citrus canker control using TRV demonstrated a remarkable 73% reduction in water volume and a 50% reduction in copper rates compared to standard applications [59]. More recent findings by Behlau et al. affirmed the efficacy of the TRV method, revealing that citrus canker control can be achieved with just 40 mL of water volume and a copper rate of 36.8 mg/m^3 [60].

Employing resistant Xcc citrus varieties is an effective means of pathogen control however, developing these varieties is both expensive and time-intensive. An environmentally friendly alternative for managing citrus canker involves utilizing endophytic bacteria as a biological control method. The utilization of endophytic bacteria in biocontrol not only enhances resistance to diseases but also presents a promising substitute for chemical control strategies [61]. This approach offers a viable and ecologically sound option for combating citrus canker. An alternative approach with potential application involves the induction of Systemic Acquired Resistance (SAR) as a means of controlling phytopathogens [62]. SAR serves as an induced defense mechanism, triggered by salicylic acid, and leads to the accumulation of pathogenicity-related proteins, bolstering the plant's resistance to various microorganisms. Molecules analogous to salicylic acid, such as isonicotinic acid (INA) and acibenzolar-S-methyl (ASM), can also induce SAR and serve as pathogen control agents. In a study on Swingle citrumelo, soil application of SAR inducers, including imidacloprid, INA, and ASM, reduced lesion numbers and altered their characteristics, resulting in smaller, darker, and less eruptive lesions. This reduction in lesions and a shift in phenotype were associated with a decrease in Xcc population, suggesting that inducing SAR response in susceptible citrus species holds promise for controlling Xcc [63]. A recent investigation into novel approaches for citrus canker disease control focused on the application of penicillin acid to citrus leaves, which showed a decrease in disease lesions indicating the potential of penicillin acid as an effective agent for Citrus Canker disease control [64].

Presently, the identification of citrus canker relies on bacterial culture methods and PCR techniques. While PCR offers rapid results, it may yield false negatives due to PCR inhibitors in samples, and culture techniques, although more reliable, are time-intensive [65, 66]. Consequently, there is a pressing need for the development of an accurate and early field detection method for citrus canker. Haji-Hashemi et al. investigated an electrochemical immunosensor targeting the effector protein PthA for citrus canker diagnosis, demonstrating high selectivity, stability, repeatability, and considerable potential for real sample applications [67]. Abdulridha et al. explored the application of remote sensing technology, employing an unmanned aerial vehicle and hyperspectral imaging, achieving high accuracy in detecting citrus canker in infected leaves during asymptomatic developmental stages [65].

3.2 Huanglongbing

At present, there are no citrus varieties displaying resistance to HLB, and no identified cure exists for this disease. Additionally, efforts to eradicate HLB from any specific area have not been successful [68]. HLB poses many challenges which include the similarity of its symptoms to those of other citrus diseases, a limited understanding of infection mechanisms, the uneven distribution of the pathogen within the host, and the inability to cultivate the bacterium *in vitro*; collectively presenting a significant hurdle for its study and thus management [30, 69–71]. Effective monitoring of psyllids in commercial areas and their control through insecticides is crucial for preventing the dissemination of HLB [72, 73] but persistent use of insecticides like imidacloprid, acetamiprid, nitenpyram, clothianidin, thiamethoxam, and cyantraniliprole, in excess, results in high production expenses, the emergence of resistant individuals, environmental pollution, and potential disruption of the ecological balance between pests and their natural adversaries, leading to outbreaks and pest resurgence [74–77]. So regular general practices like removing infected trees from the orchard can be followed. Regular inspections are crucial to identify new symptomatic plants or those overlooked in prior assessments. Once visually identified, the removal of symptomatic plants is essential to reduce inoculum sources and prevent further spread [3]. Numerous alternative approaches have been explored for managing the issue, including the employment of chemotherapy, thermotherapy, the application of antibiotics, the utilization of natural enemies of the disease vector, and the injection of defense activators into the plant trunk [75, 76, 78]. Nevertheless, all these approaches have their limitations, thereby allowing citrus to persist as a source of inoculation for HLB disease [75, 78].

HLB detection using Polymerase Chain Reaction (PCR) [79] faces challenges due to the irregular distribution of CLas, leading to potential false-negative outcomes. Additionally, the method's high cost and time-intensive sample preparation make its widespread implementation [38, 70]. Thus, no field-deployable device has been officially validated through PCR results for reliable use in diagnosing. Pagliai et al. demonstrated a potential therapeutic approach against CLas by selectively inhibiting the transcriptional activator LdtR, a key regulator of the ldtP gene encoding a transpeptidase. Through screening over 1300 small molecules, including phloretin, hexestrol, and benzbromarone, they identified compounds that hindered the binding of LdtR to DNA. *In vitro* assays revealed that these molecules induced morphological changes and osmotic stress sensitivity in CLas cells, suggesting their potential efficacy in controlling HLB [80]. Recent techniques used for detection involve Raman spectroscopy (effective in discerning HLB in citrus by comparing CLas-infected, healthy, and nutritionally deficient samples. Its high sensitivity relies on detecting molecules secreted in early infection stages, surpassing PCR's detection limits for low CLas levels [81], unique biomarkers like sec-delivered effector 1 (SDE1) (expressed highly in early-stage infected citrus tissues, systemic distribution of which in trees minimizes false negatives and the required sample collection per tree) [70] and use of desorption electrospray ionization coupled to mass spectrometer imaging (DESI-MSI) in HLB monitoring by studying various metabolites, including organic acids, phytohormones, sugars, and amino acids, associated with the plant defense system [82].

Nevertheless, replicated field studies have shown that canines can achieve over 99% accuracy in detecting CLas- and CLaf-infected plants. Preliminary findings also suggest that canines can effectively alert trained human handlers to CLam-infected plants [83]. The urgent need for early detection methods for CLas transmission, given its brief latency period, is a primary focus of current research efforts.

3.3 Citrus blast

Reducing inoculum sources of Pss involves pruning dead or diseased twigs and shoots in spring and post-rainy seasons, which effectively diminishes the incidence and severity of citrus blast and black pit disease by curbing pathogen spread. Implementing wind protection measures, such as installing windbreaks to mitigate wind-induced injuries, is recommended, considering wind is a primary infection inducer. Monitoring tree nutritional status, coupled with adhering to a well-structured fertilization schedule and regular pruning, proves essential to forestall excessive susceptible new growth in the autumn [41]. Opting for resistant or tolerant cultivars stands as an effective approach to disease control [13]. While no citrus cultivar exhibits complete resistance to blast and black pit, certain cultivars demonstrate lower susceptibility. 'Eureka', lemon cultivar displays the least susceptibility to citrus blast disease, whereas clementine cultivars such as 'MA3' and 'Cassar' tend to be less susceptible to black pit [13]. Consequently, these cultivars are recommended for replanting affected citrus orchards, particularly in cold and wet environments conducive to citrus bacteriosis development. For new plantings, it is advisable to choose bushy cultivars with relatively few thorns to minimize vulnerability to injury during severe windstorms [13, 41].

The chemical management of Pss involves a preventive use of copper compounds [84]. The application of protective copper-containing sprays, such as Bordeaux mixture or fixed copper compounds, is advised in conditions conducive to the disease. It is recommended to apply these sprays timely during the fall and early winter, before the initial rains, to safeguard leaves and shoots from Pss infections during late winter and spring [41]. While copper products prove highly effective against bacterial pathogens, their prolonged use may pose significant environmental and health risks, including the emergence of copper-resistant bacterial strains. Considering these concerns, biological control methods are strongly advocated [85, 86]. Biological control involves employing living organisms and their byproducts to restrict or suppress pathogen activities and populations [87]. Plant extracts and certain biological agents, such as *Bacillus* spp., can play a significant role in suppressing or managing citrus bacteriosis [61]. As noted by noted by Mougou et al., garlic extract and *Bacillus spp*. exhibit robust antimicrobial activity against the bacterial agents responsible for citrus blast [40] in both *in vitro* and greenhouse conditions. Additionally, a preliminary study explored the use of phage treatment for controlling bacterial blast in citrus species [88]. Consequently, biological control methods hold significant potential for pathogen management, although further research is required to assess their economic feasibility.

3.4 Citrus variegated chlorosis (CVC)

Management of CVC has traditionally involved preventive measures, primarily the eradication of symptomatic citrus trees, to diminish inoculum sources and minimize the risk of recurrent infections. While weeds and certain arboreal plants may host Xfp, their limited bacterial titer and short life substantially reduce their significance as CVC sources [89]. The accurate identification of diseased trees is crucial, demanding frequent and meticulous inspections due to the continuous emergence of new symptomatic trees resulting from ongoing infections. Under specific conditions, the health of diseased trees can be restored through drastic pruning of symptomatic branches, but this approach is successful only for trees over 2.5 years old and exhibiting symptoms on the tip of a single secondary branch [90].

A novel mitigation measure "Scion substitution" strategy has been developed based on the observed high resistance to Xfp infection in the Cravo Rangpur lime rootstock [91]. Utilizing these rootstocks, where the bacteria do not move downward below the graft union, healthy scions can be generated by removing all infected scions, even from severely affected sweet orange trees, and grafting them with healthy budwood onto these rootstocks. Unfortunately, the rootstock's immunity to CVC does not transfer to the scion. Nonetheless, the new shoots on the rootstock trunk produce fruits more rapidly and at lower costs than those on newly planted nursery trees [47]; but as there is inconsistent adoption of this technique by growers, certain areas may face elevated inoculum pressure from neighboring groves. Hence, effective control of the insect vector is vital for CVC control. Presently, this control involves the application of systemic insecticides to the trunk of younger trees designated for replanting [92].

Mauricio et al. conducted a 6-year study on the susceptibility of 264 Murcott tangor (*Citrus reticulata* Blanco × *C. sinensis* (L.) Osbeck) and Pera sweet orange (*C. sinensis* (L.) Osbeck) hybrids. Their research revealed that plants exhibiting high expression of isoflavone reductase displayed significant resistance to Xfp infection. Additionally, the study involved the assessment of 12 defense-related genes in both resistant and susceptible varieties to CVC disease. The findings indicated significantly elevated expression in resistant varieties, suggesting the involvement of these genes in conferring resistance against CVC. Among the identified resistance-related genes were those associated with disease resistance proteins (RGA2 and DRP), protein kinase activity (LRR-RK and FLS2), a nucleotide-binding protein (IFR2), and a transcription factor (b-Zip). This outcome underscores that the resistance of these citrus hybrids to Xfp is attributed not to their anatomical structure but rather to the defense mechanisms inherent in these hybrids [93]. Recently, Pereira et al. identified that the CrRAP2.2 gene in *C. reticulata* is linked to resistance against *X. fastidiosa* infection in this citrus variety. This gene shares homology with *Arabidopsis thaliana* AtRAP2.2, associated with the transcriptional factor RAP2.2 and resistance to the pathogen Botrytis cinerea in *A. thaliana* [94, 95]. The expression of the CrRAP2.2 gene was found to induce resistance to CVC in *C. sinensis*, subsequently reducing disease symptoms [95]. Developing diagnostic methods for CVC during its asymptomatic phase is crucial for early citrus removal, preventing it from becoming an inoculum source for the bacteria. In this context, Soares et al. employed Liquid Chromatography coupled with Atmospheric Pressure Chemical Ionization-Mass Spectrometry-Selected Reaction Monitoring (LC/APCI-MS-SRM) to detect CVC before observable symptoms appeared in affected trees. The study analyzed samples of *C. sinensis* grafted on *C. limonia*, revealing elevated flavonoid concentrations in leaves and coumarins in roots of symptomatic plants compared to asymptomatic and healthy plants. Based on these findings, Soares et al. suggested that the LC/APCI-MS-SRM method could effectively detect CVC before symptomatic manifestations, offering a simple and accurate diagnostic approach [96].

4. Conclusion and prospect for future

Understanding the pathogenicity mechanisms of these diseases remains an ongoing challenge. The substantial genetic variability of phytopathogens like *X. fastidiosa* and *Xanthomonas* spp., and difficulties in culturing the bacterium *Candidatus Liberibacter spp*. contribute to gaps in knowledge [14, 97–99]. Citrus canker, CVC,

HLB and citrus blast have exerted a lasting impact on the global economy for decades [38, 100]. The rapid and imperceptible transmission of these diseases poses a significant challenge for citrus growers, intensified by the absence of effective treatments, ultimately resulting in the demise of infected plants [3, 93, 98, 101]. In response to these challenges, current research efforts are focused on unraveling the mechanisms of infection during the initial stages of the diseases, employing analytical approaches [70, 82, 96, 102]. The advancement of biochemical techniques and analytical platforms is anticipated to enhance our understanding of infection and pathogenicity mechanisms, potentially paving the way for the discovery of effective strategies for diagnosing and controlling major citrus diseases.

Conflict of interest

The authors declare no conflict of interest.

Author details

Aditya Kukreti[1*] and Namburi Karunakar Reddy[2]

1 Department of Plant Pathology, Indira Gandhi Krishi Vishwavidyalaya, Raipur, Chhattisgarh, India

2 Department of Plant Pathology, University of Agricultural Sciences, GKVK, Bengaluru, India

*Address all correspondence to: akukreti22@gmail.com

References

[1] Caserta R, Teixeira-Silva NS, Granato LM, Dorta SO, Rodrigues CM, Mitre LK, et al. Citrus biotechnology: What has been done to improve disease resistance in such an important crop? Biotechnology Research and Innovation. 2020;**3**:95-109. DOI: 10.1016/j.biori.2019.12.004

[2] Mendonca LBP, Zambolim L, Badel JL. Bacterial citrus diseases: Major threats and recent progress. Journal of Bacteriology and Mycology. 2017;**5**(4):340-350. DOI: 10.15406/jbmoa.2017.05.00143

[3] Gabriel D, Gottwald TR, Lopes SA, Wulff NA. Bacterial pathogens of citrus: Citrus canker, citrus variegated chlorosis and Huanglongbing. Genus Citrus. 2020;**2020**:371-389. DOI: 10.1016/B978-0-12-812163-4.00018-8

[4] Raddadi N, Gonella E, Camerota C, Pizzinat A, Tedeschi R, Crotti E, et al. 'Candidatus *Liberibacter europaeus*' sp. nov. that is associated with and transmitted by the psyllid *Cacopsylla pyri* apparently behaves as an endophyte rather than a pathogen. Environmental Microbiology. 2011;**13**(2):414-426. DOI: 10.1111/j.1462-2920.2010.02347.x

[5] European and Mediterranean Plant Protection Organization (EPPO). 2020. EPPO global database. Available from: https://gd.eppo.int.

[6] Sharma SK, Sharma RR. Citrus canker approaching century: A review. Tree and Forestry Science and Biotechnology. 2009;**2**(2):54-56

[7] Alves MN, Lopes SA, Raiol-Junior LL, Wulff NA, Girardi EA, Ollitrault P, et al. Resistance to 'Candidatus *Liberibacter asiaticus*,' the huanglongbing associated bacterium, in sexually and/or graft-compatible citrus relatives. Frontiers in Plant Science. 2021;**11**:617664

[8] Mirik M, Baloglu S, Aysan Y, Cetinkaya-Yildiz R, Kusek M, Sahin F. First outbreak and occurrence of citrus blast disease, caused by *Pseudomonas syringae* pv. syringae, on orange and mandarin trees in Turkey. Plant Pathology. 2005;**54**:238-238

[9] Beiki F, Busquets A, Gomila M, Rahimian H, Lalucat J, Garcia-Valdes E. New *Pseudomonas* spp. are pathogenic to citrus. PLoS One. 2016;**11**:1-16

[10] Ivanović Z, Perovic T, Popovic T, Blagojevic J, Trkulja N, Hrncic S. Characterization of *Pseudomonas syringae* pv. *syringae*, causal agent of citrus blast of mandarin in Montenegro. Plant Pathology Journal. 2017;**33**:21-33

[11] Abdellatif E, Kaluzna M, Helali F, Cherif M, Janse JD, Rhouma A. First report of citrus bacterial blast and citrus black pit caused by *Pseudomonas syringae* pv. Syringae in Tunisia. New Disease Reports. 2015;**32**:35

[12] Mougou I. Citrus blast and black pit disease: A review. DYSONA-Life Science. 2022;**3**(1):1-6

[13] Mougou I, Boughalleb-M'hamdi N. Detection, survival, and source of inoculum of *Pseudomonas syringe* pv. syringae from weeds and plant debris in relation to epidemiology of bacterial citrus blast and black pit in Tunisia. British Microbiology Research Journal. 2016;**16**:1-10

[14] Nunney L, Schuenzel EL, Scally M, Bcromley RE, Stouthamer R. Large-scarle intersubspecific recombination in

the plant pathogenic bacterium *Xylella fastidiosa* is associated with the host shift to mulberry. Applied and Environmental Microbiology. 2014;**80**(10):3025-3033. DOI: 10.1128/AEM.04112-13

[15] Contreras JDV. Pecosita ou falsa mancha grasienta na Argentina. Laranja e Cia. 1992;**31**:6

[16] Fadel AL, Stuchi ES, de Carvalho SA, Federici MT, Della Coletta-Filho H. Navelina ISA 315: A cultivar resistant to citrus variegated chlorosis. Crop Protection. 2014;**64**:115-121

[17] Niza B, Coletta-Filho H, Merfa M, Takita M, de Souza A. Differential colonization patterns of *Xylella fastidiosa* infecting citrus genotypes. Plant Pathology. 2015;**64**:1259-1269

[18] Fonseca NP, Patané JSL, Varani AM, Felestrino ÉB, Caneschi WL, Sanchez AM, et al. Analyses of seven new genomes of *Xanthomonas citri* pv. Aurantifolii strains, causative agents of citrus canker B and C, show reduced repertoire of pathogenicity-related genes. Frontiers in Microbiology. 2019;**10**:2361. DOI: 10.3389/fmicb.2019.02361

[19] Bahadur A, Srivastava JN. Integrated management of citrus canker disease (Xanthomonas axonopodis pv. Citri). In: Srivastava JN, Singh AK, editors. Diseases of Horticultural Crops: Diagnosis and Management. Boca Raton: Taylor & Francis Group; 2022. pp. 170-183

[20] Li J, Wang N. Foliar application of biofilm formation-inhibiting compounds enhances control of citrus canker caused by *Xanthomonas citri* subsp. citri. Phytopathology. 2014;**104**(2):134-142. DOI: 10.1094/PHYTO-04-13-0100-R

[21] Pontes JGM, Fernandes LS, dos Santos RV, Tasic L, Fill TP. Virulence factors in the phytopathogen–host interactions: An overview. Journal of Agricultural and Food Chemistry. 2020;**68**(29):7555-7570. DOI: 10.1021/acs.jafc.0c02389

[22] Brunings AM, Gabriel DW. *Xanthomonas citri*: Breaking the surface. Molecular Plant Pathology. 2003;**4**(3):141-157. DOI: 10.1046/j.1364-3703.2003.00163.x

[23] Gabriel DW, Duane YP, Ramadugu C. The Molecular Mechanism of Citrus Canker Pathogenicity and a Gene Engineering Approach to Control. Orlando, FL: International Society of Citriculture; 2000. p. 51

[24] Das AK. Citrus canker – A review. Journal of Applied Horticultre. 2003;**5**(1):52-60. DOI: 10.37855/jah.2003.v05i01.15

[25] Graham JH, Gottwald TR. Research perspectives on eradication of citrus bacterial diseases in Florida. Plant Disease. 1991;**75**:1193-1200

[26] Gottwald TR, Graham JH, Schubert TS. Citrus canker: The pathogen and its impact. Plant Health Progress. 2002;**3**:15. Available from: www.plantmanagementnetwork.org/pub/php/review/citruscanker

[27] Gottwald TR, Irey M. Post-hurricane analysis of citrus canker II: Predictive model estimation of disease spread and area potentially impacted by various eradication protocols following catastrophic weather events. Plant Health Progress. 2007;**8**(1):22. DOI: 10.1094/PHP-2007-0405- 01-RS April 2007

[28] Gottwald TR, Graham JH, Schubert TS. An epidemiological analysis of the spread of citrus canker in urban Miami, Florida, and synergistic interaction with the Asian citrus leafminer. Fruits. 1997;**52**:383-390

[29] Gottwald TR, Bassanezi RB, Amorim L, Bergamin-Filho A. Spatial pattern analysis of citrus canker infected plantings in São Paulo Brazil and augmentation of infection elicited by the Asian Leafminer. Phytopathology. 2007;**97**:674-683

[30] Bové J. Huanglongbing: A destructive, newly-emerging, century-old disease of citrus. Journal of Plant Pathology. 2006;**88**(1):7-37. DOI: 10.4454/jpp.v88il.828

[31] da Graça JV. Citrus greening disease. Annual Review of Phytopathology. 1991;**29**:109-136

[32] Jagoueix S, Bové JM, Garnier M. The phloem-limited bacterium of greening disease of citrus is a member of the α subdivision of the proteobacteria. International Journal of Systematic Bacteriology. 1994;**44**:379-386

[33] Garnier M, Bové JM, Sanders GM, Korsten L, Le Roux HF. Presence of "Candidatus liberibacter africanus" in the western cape province of South Africa. In: Proceedings of 14th Conference of International Organization of Citrus Virologists, 13-18 September 1998. Campinas, São Paulo State, Brazil: International Organization of Citrus Virologists; 2000. pp. 369-372. DOI: 10.5070/C52SZ5Q3W7

[34] Teixeira DC, Danet JL, Eveillard S, Martins EC, Jesus Junior WC, Yamamoto PT, et al. Citrus huanglongbing in São Paulo state, Brazil: PCR detection of the 'Candidatus' liberibacter species associated with the disease. Molecular and Cellular Probes. 2005;**19**(3):173-179. DOI: 10.1016/j.mcp.2004.11.002

[35] Ammar E-D, George J, Sturgeon K, Stelinski LL, Shatters RG. Asian citrus psyllid adults inoculate huanglongbing bacterium more efficiently than nymphs when this bacterium is acquired by early instar nymphs. Scientific Reports. 2020;**10**:18244. DOI: 10.1038/s41598-020-75249-5

[36] Lee JA, Halbert SE, Dawson WO, Robertson CJ, Keesling JE, Singer BH. Asymptomatic spread of huanglongbing and implications for disease control. Proceedings of the National Academy of Sciences of the United States of America. 2015;**112**:7605-7610

[37] Halbert SE, Manjunath KL. Asian citrus psyllids (sternorrhyncha: Psyllidae) and citrus greening disease of citrus: A literature review and assessment of risk in Florida. Florida Entomologist. 2004;**87**(3):330-353. DOI: 10.1653/0015-4040(2004)087[0330:acpspa]2.0.co;2

[38] Wetterich CB, Neves RFO, Belasque J, Ehsani R, Marcassa LG. Detection of Huanglongbing in Florida using fluorescence imaging spectroscopy and machine-learning methods. Journal of Applied Optics. 2017;**56**(1):15-23. DOI: 10.1364/AO.56.000015

[39] Lopes SA, Bertolini E, Frare GF, Martins EC, Wulff NA, Teixeira DC, et al. Graft transmission efficiencies and multiplication of "Candidatus *Liberibacter americanus*" and "Ca. *Liberibacter asiaticus*" in citrus plants. Phytopathology. 2009;**99**:301-306

[40] Mougou I, Boughalleb-M'hamdi N. Differential susceptibility of citrus cultivars toward blast and black pit in Tunisia caused by *Pseudomonas syringae* pv. syringae. European Journal of Biotechnology and Bioscience. 2016;**4**:17-24

[41] Ferguson L, Grafton-Cardwell EE. Citrus Production Manual. Oakland,

University of California: UCANR Publications; 2014. p. 433

[42] Canfield ML, Baca S, Moore LW. Isolation of *Pseudomonas syringae* from 40 cultivars of diseased woody plants with tip dieback in Pacific Northwest nurseries. Plant Disease. 1986;**70**:647-650

[43] Simpson AJG, Reinach FC, Arruda P, Abreu FA, Acencio M, Alvarenga R, et al. The genome sequence of the plant pathogen *Xylella fastidiosa*. Nature. 2000;**406**:151-157. DOI: 10.1038/35018003

[44] Newman KL, Almeida RPP, Purcell AH, Lindow SE. Use of a green fluorescent strain for analysis of *Xylella fastidiosa* colonization of *Vitis vinifera*. Applied and Environmental Microbiology. 2003;**69**(12):7319-7327. DOI: 10.1128/AEM.69.12.7319-7327.2003

[45] Roper MC, Greve LC, Warren JG, Labavitch JM, Kirkpatrick BC. *Xylella fastidiosa* requires polygalacturonase for colonization and pathogenicity in *Vitis vinifera* grapevines. Molecular Plant-Microbe Interactions. 2006;**20**:411-419

[46] Queiróz-Voltan RB, Paradela FO. Caracterização de estruturas anatômicas de citros infectados com *Xylella fastidiosa*. Laranja. 1999;**20**:55-76

[47] Garcia AL, Torres SCZ, Heredia M, Lopes SA. Citrus responses to *Xylella fastidiosa* infection. Plant Disease. 2012;**96**:1245-1249

[48] Landa BB, Saponari M, Feitosa-Junior OR, Giampetruzzi A, Vieira FJD, Mor E, et al. *Xylella fastidiosa's* relationships: The bacterium, the host plants, and the plant microbiome. New Phytologist. 2022;**234**:1598-1605. DOI: 10.1111/nph.18089

[49] He CX, Li WB, Ayres AJ, Hartung JS, Miranda VS, Teixeira DC. Distribution of *Xylella fastidiosa* in citrus rootstocks and transmission of citrus variegated chlorosis between sweet orange plants through natural root grafts. Plant Disease. 2000;**84**(6):622-626. DOI: 10.1094/PDIS.2000.84.6.622

[50] Lopes JRS, Krugner R. Transmission ecology and epidemiology of the citrus variegated chlorosis strain of Xylella fastidiosa. In: Brown JK, editor. Vector-Mediated Transmission of Plant Pathogens. American Phytopathological Society Press; 2016. pp. 195-208. DOI: 10.1094/9780890545355.014

[51] Marcus IM, White D, Backus EA, Walker SL, Roper MC. Fluid dynamic simulations at the interface of the blue-green sharpshooter functional foregut and grapevine xylem sap with implications for transmission of *Xylella fastidiosa*. PLoS One. 2022;**17**:e0265762. DOI: 10.1371/journal.pone.0265762

[52] Laranjeira FF, Palazzo D. Danos qualitativos à produção de laranja 'Natal' causados pela clorose variegada dos citros. Laranja. 1999;**20**:77-91

[53] Ayres AJ. Intensidade da clorose variegada dos citros em pomares comerciais de laranja do estado de São Paulo e sul do triângulo mineiro [thesis]. Brasil: Universidade Estadual Paulista, Faculdade de Ciências Agrárias e Veterinárias, Jaboticabal; 2000

[54] Leite RP Jr, Mohan SK, Pereira ALG, Campacci CA. Integrated control of citrus canker—Effect of genetic resistance and application of bactericides. Fitopatologia-Brasileira. 1987;**12**:257-263

[55] Graham JH, Gottwald TR, Leite RP Jr. Prospects for control of citrus canker with novel chemical compounds. Proceedings of the Florida State Horticultural Society. 2006;**119**:83-88

[56] Behlau F, Jones JB, Myers ME, Graham JH. Monitoring for resistant populations of *Xanthomonas citri* subsp. citri and epiphytic bacteria on citrus trees treated with copper or streptomycin using a new semi-selective medium. European Journal of Plant Pathology. 2012;**132**:259-270

[57] Pernezny K, Nagata R, Havranek N, Sanchez J. Comparison of two culture media for determination of the copper resistance of Xanthomonas strains and their usefulness for prediction of control with copper bactericides. Crop Protection. 2008;**27**(2):256-262. DOI: 10.1016/j.cropro.2007.05.012

[58] Byers RE, Hickey KD, Hill CH. Base gallonage per acre. Va. Fruit. 1971;**60**(8):19-23

[59] da Silva Scapin M, Behlau F, Scandelai LHM, Fernandes RS, Junior GJS, Ramos HH. Tree-row-volume-based sprays of copper bactericide for control of citrus canker. Crop Protection. 2015;**77**:119-126

[60] Behlau F, Lanza FE, da Scapin MS, Scadelai LHM, Silva Júnior GJ. Spray volume and rate based on the tree row volume for a sustainable use of copper in the control of citrus canker. Plant Disease. 2021;**105**:183-192. DOI: 10.1094/PDIS-12-19-2673-RE

[61] Islam MN, Ali MS, Choi SJ, Hyun JW, Baek KH. Biocontrol of citrus canker disease caused by *Xanthomonas citri* subsp. citri using an endophytic *Bacillus thuringiensis*. Plant Pathology Journal. 2019;**35**(5):486-497. DOI: 10.5423/PPJ.OA.03.2019.0060

[62] Walters DR, Ratsep J, Havis ND. Controlling crop diseases using induced resistance: Challenges for the future. Journal of Experimental Botany. 2013;**64**:1263-1280. DOI: 10.1093/jxb/ert026

[63] Francis MI, Redondo A, Burns JK, Graham JH. Soil application of imidacloprid and related SAR-inducing compounds produces effective and persistent control of citrus canker. European Journal of Plant Pathology. 2009;**124**:283-292. DOI: 10.1007/s10658-008-9415-x

[64] Vieira G, Khalil ZG, Capon RJ, Sette LD, Ferreira H, Sass DC. Isolation and agricultural potential of penicillic acid against citrus canker. Journal of Applied Microbiology. 2022;**132**:3081-3088. DOI: 10.1111/jam.15413

[65] Abdulridha J, Batuman O, Ampatzidis Y. UAV-based remote sensing technique to detect citrus canker disease utilizing hyperspectral imaging and machine learning. Remote Sensing. 2019;**11**(1):1373. DOI: 10.3390/rs11111373

[66] Naqvi SAH, Wang J, Malik MT, Umar UUD, Rehman AU, Hasnain A, et al. Citrus canker—Distribution, taxonomy, epidemiology, disease cycle, pathogen biology, detection, and management: A critical review and future research agenda. Agronomy. 2022;**12**:1075. DOI: 10.3390/agronomy12051075

[67] Haji-Hashemi H, Norouzi P, Safarnejad MR, Larijani B, Habibi MM, Raeisi H, et al. Sensitive electrochemical immunosensor for citrus bacterial disease detection using fast Fourier transformation square-wave voltammetry method. Journal of Electroanalytical Chemistry. 2018;**820**:111-117. DOI: 10.1016/j.jelechem.2018.04.062

[68] Gottwald TR. Current epidemiological understanding of citrus Huanglongbing. Annual Review of Phytopathology. 2010;**48**:119-139

[69] Bové JM, Garnier M. Phloem- and xylem-restricted plant pathogenic bacteria. Plant Science. 2003;**163**(6):1083-1098. DOI: 10.1016/S0168-9452(02)00276-5

[70] Tran TT, Clark K, Ma W, Mulchandani A. Detection of a secreted protein biomarker for citrus Huanglongbing using a single-walled carbon nanotubesbased chemiresistive biosensor. Biosensors & Bioelectronics. 2020;**147**(111766):111766-111769. DOI: 10.1016/j.bios.2019.111766

[71] Wang N. The citrus Huanglongbing crisis and potential solutions. Molecular Plant. 2019;**12**:607-609. DOI: 10.1016/j.molp.2019.03.008

[72] Bassanezi RB, Montesino LH, Gimenes-Fernandes N, Yamamoto PT, Gottwald TR, Amorim L, et al. Efficacy of area-wide inoculum reduction and vector control on temporal progress of huanglongbing in young sweet orange plantings. Plant Disease. 2013;**97**:789-796

[73] Miranda MP, Santos FL, Bassanezi RB, Montesino LH, Barbosa JC, Sétamou M. Monitoring methods for *Diaphorina citri* Kuwayama (Hemiptera: Liviidae) on citrus groves with different insecticide application programmes. Journal of Applied Entomology. 2017;**142**:1-8

[74] Naem A, Freed S, Jin FL, Akmal M, Mehmood M. Monitoring of insecticide resistance in *Diaphorina citri* Kuwayama (Hemiptera: Psyllidae) from citrus groves of Punjab, Pakistan. Crop Protection. 2016;**86**:62-68. DOI: 10.1016/j.cropro.2016.04.010

[75] Dorta SO, Machado MA, Freitas-Astúa J. Desenvolvimento de estratégias alternativas visando ao controle do Huanglongbing. Citrus R&T. 2019;**40**:e1045. DOI: 10.4322/crt.17519

[76] Li J, Pang Z, Duan S, Lee D, Kolbasov VG, Wang N. The in planta effective concentration of oxytetracycline against 'Candidatus *Liberibacter asiaticus*' for suppression of citrus Huanglongbing. Phytopathology. 2019;**109**(12):2046-2054. DOI: 10.1094/PHYTO-06-19-0198-R

[77] Uthman QO, Atta AA, Kadyampakeni DM, Qureshi JA, Morgan KT, Nkedi-Kizza P. Integrated water, nutrient, and pesticide management of Huanglongbing-affected sweet oranges on Florida sandy soils - a review. Plants. 2022;**11**:1850. DOI: 10.3390/plants11141850

[78] Zhang M, Yang C, Powell CA, Avery PB, Wang J, Huang Y, et al. Field evaluation of integrated management for mitigating citrus Huanglongbing in Florida. Frontiers in Plant Science. 2019;**9**:1890. DOI: 10.3389/fpls.2018.01890

[79] Innis MA, Gelfand DH, Sninsky JJ, White TJ. PCR Protocols: A Guide to Methods and Applications. San Diego: Academic Press; 1990

[80] Pagliai FA, Gardner CL, Bojilova L, Sarnegrim A, Tamayo C, et al. The transcriptional activator LdtR from 'Candidatus *Liberibacter asiaticus*' mediates osmotic stress tolerance. PLoS Pathogens. 2014;**10**(4):e1004101

[81] Sanchez L, Pant S, Xing Z, Mandadi K, Kurouski D. Rapid and noninvasive diagnostics of Huanglongbing and nutrient deficits on citrus trees with a handheld Raman spectrometer. Analytical and Bioanalytical Chemistry. 2019;**411**:3125-3133. DOI: 10.1007/s00216-019-01776-4

[82] Pontes JGM, Vendramini PH, Fernandes LS, de Souza FH, Pilau EJ, Eberlin MN, et al. Mass spectrometry imaging as a potential technique for diagnostic of Huanglongbing disease using fast and simple sample preparation. Scientific Reports. 2020;**10**:13457. DOI: 10.1038/s41598-020-70385-4

[83] Gottwald T, Poole G, Taylor E, Hall D, Hartung J, Bartels D, et al. Canine assisted early detection of HLB. Journal of Citrus Pathology. 2017;**4**(1):16

[84] Montesinos E, Lopez MM. Métodos de control de las bacteriosis. In: Llacer G, Lopez MM, Trapero T, Bello A, editors. Patologia Vegetal. Vol. Tomo 1. Espana: Societad Espanola de Fitopatologia; 1996. pp. 653-678

[85] Fravel DR. Commercialization and implementation of biocontrol. Annual Review of Phytopathology. 2005;**43**:337-359

[86] Stewart A, Hill R, Stark C. Desktop Evaluation on Commercially Available Microbial-Based Products for Control or Suppression of Pseudomonas syringae Pv. Actinidiae. Lincoln, New Zealand: Bio-Protection Research Centre, Lincoln University; 2011. p. 33

[87] Dekker M. Biological Control of Crop Diseases. Vol. 2002. New York: CRC Press; 2002. p. 468

[88] Pinheiro LAM, Pereira C, Frazão C, Balcão VM, Almeida A. Efficiency of phage φ6 for biocontrol of *Pseudomonas syringae* pv. syringae: An *in vitro* preliminary study. Microorganisms. 2019;**7**(9):286

[89] Lopes SA, Marcussi S, Torres SCZ, Souza V, Fagan C, França SC. Weeds as alternative hosts of the citrus, coffee, and plum strains of *Xylella fastidiosa* in Brazil. Plant Disease. 2003;**87**:544-549

[90] Rodas V. Convivência com a clorose variegada dos citros. Laranja. 1994;**15**:129-134

[91] Lopes SA. A new strategy to control citrus variegated chlorosis disease. Plant Disease. 2019;**104**(1):239-245. DOI: 10.1094/PDIS-02-19-0302-RE

[92] Yamamoto PT. Controle de insetos vetores de bactérias causadoras em doenças dos citros. In: Manejo integrado de pragas dos citros. Fundo de Defesa da Citricultura (Fundecitrus), Araraquara, Brazil: esalq/usp piracicaba sp, APS Press; 2008. pp. 237-260

[93] Mauricio FN, Soratto TAT, Diogo JA, Boscariol-Camargo RL, De Souza AA, Coletta-Filho HD, et al. Analysis of defense-related gene expression in citrus hybrids infected by *Xylella fastidiosa*. Phytopathology. 2019;**109**:301-306. DOI: 10.1094/phyto-09-18-0366-f

[94] Zhao Y, Wei T, Yin KQ, Chen Z, Gu H, Qu LJ, et al. Arabidopsis RAP2.2 plays an important role in plant resistance to *Botrytis cinerea* and ethylene responses. The New Phytologist. 2012;**195**:450-460. DOI: 10.1111/j.1469-8137.2012.04160.x

[95] Pereira W, Takita M, Souza MMA, de Souza A. Citrus reticulata CrRAP2.2 transcriptional factor similar functions to the *Arabidopsis homolog* and increases resistance to *Xylella fastidiosa*. Molecular Plant-Microbe Interactions. 2020;**33**(3):519-527. DOI: 10.1094/MPMI-10-19-0298-R

[96] Soares MS, da Silva DF, Amaral JC, da Silva MM, Forim MR, Rodrigues Filho E, et al. Rapid differentiation of graft *Citrus sinensis* with and without *Xylella fastidiosa* infection by mass spectrometry. Rapid Communications in Mass Spectrometry. 2020;**34**:e8745. DOI: 10.1002/rcm.8745

[97] Carvalho FMS, Caramori LPC, Júnior RPL. Genetic diversity of *Xanthomonas axonopodis* pv. citri based on plasmid profile and pulsed field gel electrophoresis. Genetics and Molecular Biology. 2005;**28**(3):446-451. DOI: 10.1590/S1415-47572005000300020

[98] Frare GF, Silva Junior GJ, Lanza FE, Bassanezi RB, Ramires TG, Amorim L. Sweet orange fruit age and inoculum concentration affect the expression of citrus black spot symptoms. Plant Disease. 2019;**103**(5):913-921. DOI: 10.1094/PDIS-03-18-0492-RE

[99] Merfa MV, López EP, Naranjo E, Jain M, Gabriel DW, De La Fuente L. Progress and obstacles in culturing 'Candidatus *Liberibacter asiaticus*', the bacterium associated with huanglongbing. Phytopathology. 2019;**109**(7):1092-1101. DOI: 10.1094/PHYTO-02-19-0051-RVW

[100] Bassanezi RB, Silva Junior GS, Feichtenberger E, Belasque Junior J, Behlau F, Wulff N. A. Doenças dos Citros. In: Amorim L, Rezende JAM, Filho AB, Camargo LEA, editors. Manual de fitopatologia: Doenças das plantas cultivadas. Ouro Fino: Agronômica Ceres; 2016. pp. 271-306

[101] Yan Q, Wang N. High-throughput screening and analysis of genes of *Xanthomonas citri* subsp. citri involved in citrus canker symptom development. Molecular Plant-Microbe Interactions. 2012;**1**:69-84. DOI: 10.1094/mpmi-05-11-0121

[102] Çetiner H. Citrus disease detection and classification using based on convolution deep neural network. Microprocessors and Microsystems. 2022;**95**:104687. DOI: 10.1016/j.micpro.2022.104687